DINGO

For Abby and Owen Anderson

AUSTRALIAN NATURAL HISTORY SERIES

DINGO

BRAD PURCELL

CSIRO

PUBLISHING

National Library of Australia Cataloguing-in-Publication entry

Purcell, Brad
Dingo / by Brad Purcell.

9780643096936 (pbk.)

Australian natural history series.

Includes bibliographical references and index.

Dingo – Australia
Dingo – Behavior – Genetic aspects – Australia.
Dingo – Effect of habitat modification on – Australia.
599.772

Published by
CSIRO PUBLISHING
36 Gardiner Road, Clayton VIC 3168
Private Bag 10, Clayton South VIC 3169
Australia
Telephone: [+613] 9545 8555
Local call: 1300 788 000 (Australia only)
Fax: +61 3 9662 7555
Email: csiropublishing@csiro.au
Website: www.publishing.csiro.au

Front cover image: Jon Reid
All images by Brad Purcell unless stated otherwise

Set in 10.5/14 Sabon
Edited by Janet Walker
Cover and text design by James Kelly
Typeset by Oryx Publishing
Printed by Ingram Lightning Source

CSIRO PUBLISHING publishes and distributes scientific, technical and health science books and journals from Australia to a worldwide audience and conducts these activities autonomously from the research activities of the Commonwealth Scientific and Industrial Research Organisation (CSIRO).
The views expressed in this publication are those of the author(s) and do not necessarily represent those of, and should not be attributed to, the publisher or CSIRO.

Feb26_RP_ILS

CONTENTS

A dingo walks behind a hidden trap, in front of a hidden camera, and appears distracted by something at the nearby waterhole.

PREFACE

'No dog's allowed in here, son!' said the storekeeper to Roland
Breckwoldt in a supermarket in Port Augusta.

'That's not a dog,' claimed Roland in reply … 'it's a dingo!'

'That's okay then.'

Such is life for this Australian icon. Friend and foe. We want to keep them
… but we also want to kill them. We want to hear their lonesome chorus in
the evening, but we want to hear the trapper say, 'I killed that dog that was
pestering your sheep.'

Roland Breckwoldt was the first to publish the science and the
speculation about both sides of the argument. After all, the dingo that killed
the sheep deserves defence since it was possibly acting much like the sheep
farmer and trying to provide food for its family. It is this competition that
drives the controversial culture about dingoes, and it is only we humans
that can make the choice to kill dingoes or to conserve them.

Two questions: 1) 'did the dingo take the Chamberlains' baby?' and
2) 'are there any pure dingoes left?' are the most commonly asked by people
when they find out I research dingoes. It is unfortunate, however, that after
40–50 years of dingo research in Australia those answers are all people seek,
and that a simple 'yes' or 'no' is unavailable for both questions. I wasn't
born until exactly one year after the baby incident had hit the newsstands
on 18 August 1980 so I can only provide comment using available literature
about the event. The most plausible explanation for the first question,
however, is that a wild dingo had become habituated in the area due to the
frequency of human visitors. Azaria Chamberlain, the baby, was possibly
murmuring or crying in the tent and one of the local dingoes seized her in
the same way that they would if they found a lone or injured joey. Aborigines
may have had similar experiences with thylacines and dingoes over
millennia, thus creating the rule for their children to carry a firestick with
them when they left the campfire at night. Apparently they would tell their
kids that it would protect them from evil spirits, and in totemistic religions,
evil spirits may take the form of an animal, much like the dingo.

Question two, however, has to be answered rhetorically: 'What *is* a
"pure" dingo?'

Ninety per cent of the time, a pure dingo apparently is a sandy coloured animal with white paws and a white tail tip. The other 10% of the time the questioners either listen with interest whilst I explain the problems when using coat colour, skull measurements or genetics to define 'purity', or they agree with my question: 'yeah ... what is "pure"?' Adolph Hitler once attempted to create a pure race of humans based on the colour of their hair and eyes, and this eventually helped to cause World War II. So what basis do humans have to define any animals as pure? The most ironic part about Hitler's definition of pure is that the pure gene for blue eyes was a genetic mutation that occurred between 5000–6000 years ago. Does that make 'purity' a genetic mutation? If it does, is not the concept of purity an illusion?

This book has been written to provide a new understanding of the dingo, Australia's wild dog. Records of dingoes prior to those referred to in this book were generally limited, anecdotal and potentially biased diary entries from European settlers that have remained influential in dingo and wild dog management programs. Dingoes are trapped between the stature of being: 1) an infamous pest animal; 2) a spiritual totemic creature; and 3) a potential keystone species and tool to reverse some impacts of European colonisation on Australian fauna and flora. These three points encapsulate the controversial existence of dingoes. If the dingoes in question are causing problems, then they are referred to as wild dogs and have to be controlled (the politically correct way to say killed or culled) under Australian legislation. Alternatively if the wild dogs or dingoes in question are useful or hail from an iconic stature, then they are referred to as dingoes and afforded a level of protection by legislation and the public. Since only Dr Doolittle and Dr Harry Cooper can talk to the animals, then it is all the more important for society at large to have an understanding of the dingo because we can't expect them to understand us.

Ultimately Roland summarised the role of the dingo most aptly at the conclusion of his foreword:

> 'Perhaps knowing and accepting the dingo is but part of reaching a much larger understanding of our place in the Australian environment.'

I hope that this book can help to do just that: show what the dingo is or has become, and show what we can learn from its story.

ACKNOWLEDGEMENTS

Many people who assisted me during my dingo research are deserving of special thanks. First of all, Professor Robert Claude Mulley. Talk about philosophy. Rob was my principal supervisor during my doctoral research and apart from finding in him an arsenal of industry experience I also found a strong and supportive mentor and friend. It doesn't get any better than that! Co-supervisors Robert Close from the University of Western Sydney and Peter Fleming from the Vertebrate Pest Research Unit at Orange also imparted constructive criticism of the highest quality. Their enthusiasm and support has been endless and it was an honour to work with them.

The University of Western Sydney deserves many thanks for the research opportunities, intercontinental experiences and logistical support to write this book. My PhD project originally was envisaged in 2004 as a flagship project for the then new animal science degree and this book is credit to the foresight at the time, of the Dean of the College of Health and Science, Professor Mick Wilson and the Head of School, Robert Mulley. Thanks also to Professor John Bartlett, technical officer Sue Cusbert and the friendly and supportive staff and postgraduate colleagues, especially Jack Pascoe.

Research of this nature could never work without collaboration from stakeholders. Therefore I am grateful to the NSW Parks and Wildlife Division of the Department of Environment, Climate Change and Water, the Sydney Catchment Authority and the Cumberland Livestock Health and Pest Authority for funding and in-kind contributions including field accommodation, aerial services and boat services. The enlightened perspectives and enthusiasm of Andrew Glover, the man who conceived the idea for the dingo project in the Blue Mountains, became the backbone of its stability. Duncan Scott-Lawson, Andrew Simson, Glenn Meade, Brian Waldron, Kim de Govrik, Chris Banffy, Loretta Gallen, Steven Mills and Geoff Ross are just few of the staff I wish to thank personally. Thanks for the valiant efforts of dog trappers Bill Morris, Mick Davis and Andrew McDougall.

Many thanks to Barbara Triggs (Dead Finish), Justyna Paplinska and George Sofrinidi (Genetic Technologies), Alan Wilton (University of New South Wales), Oleg Nicetic and Michael Dingley (University of Western Sydney) and to Ben Allen (South Australian Arid Lands Natural Resource

Management Board) for scat, genetic, statistic, skull and movement analyses. There were many volunteers who assisted me with field work, too numerous to name individually here, so I thank you as a group.

I am also indebted for the constant support of family and friends. My parents, Bryan and Vicki Purcell, my sister Kristy Anderson, my brother Simon Purcell and my brother-in-law Aran Anderson were always encouraging. I wish to thank Kerrie McGuigan for her support and also her parents Wendy and Anthony McGuigan. My gratitude is extended to my friends, particularly Stephen Wigmore for reviewing a draft of this book, Josh Edwards and Glenn Purcell for assisting me with some illustrations and to Peter Jamieson for providing accommodation while I wrote the manuscript. Thanks to Shannon Plummer, Aran Anderson, Bryan Purcell, Rob Mulley, Lee Parker and Nick Alexander for providing outstanding photographs.

Finally I thank Nick Alexander and CSIRO for approaching me to write this book and supporting me through the experience. Thanks also to Professor Chris Dickman for his support in my chosen career.

Brad Purcell

1

INTRODUCTION

Australia is a continent of contrasts with extreme weather patterns and many varied landscapes, ecosystems and animals. The dingo *Canis lupus dingo* is one of few Australian species to inhabit the entire mainland, because others are limited in their ability to adapt to the different environments. Unlike Australian marsupials, however, dingoes are a placental mammal and a canid, and canid species are landscape specialists. Following the introduction of the dingo to Australia 4000–5000 years ago, canids were the first terrestrial carnivore to be unsurpassed in world distribution. Like most large carnivores, dingoes sometimes come into conflict with humans and managing dingoes can often be controversial.

Large carnivorous species are probably the most revered in all of nature. In ancient human civilisations, some carnivores were seen as spiritual totems, such as the gray wolf *Canis lupus lupus* by American Indians, and the dingo by Australian Aborigines. In modern European civilisations, however, these two species in particular were portrayed in fables such as *Little Red Riding Hood* and *Wombat Stew* as murderous villains or as cunning impostors respectively. The fear that carnivores instil in their prey and in their competitors has created numerous defensive behaviours throughout evolution. Although only one-third the weight of its ancestor, the 60 kg gray wolf, the dingo is not exempt from being feared as a deadly predator. Kangaroos form large mobs where sentinel animals appear to position themselves at vantage points around the mob to keep watch for any terrestrial predators like the dingo. Competing species such as humans, have invented methods as simple as fences or as elaborate as poisons to not only deter but to destroy other carnivores.

Coming to understand an animal like the Australian dingo is more challenging than simply reading about them or seeing them in a zoo. Dingoes after all are an iconic Australian mammal, similar to kangaroos and koalas. They may be depicted on postcards and stamps, on instant lottery scratchies, in brochures, children's books and even on television and in movies. In contrast, dingoes are persecuted daily by many present-day Australians for both being a higher order predator and for killing introduced domestic herbivores fenced by sparse wire adjacent to dingo habitat (see Figure 1.1). If dingoes are not being shot at, trapped or poisoned, they are being caged

Figure 1.1 Leghold traps are usually set in the humus layer of soil under a scent post. Dingoes unknowingly stand on the treadle when they move in to refresh their scent and are shot by the trapper. Image: Lee Parker

or isolated in zoos and breeding sanctuaries. Millions of dollars are spent annually to control the effects of dingoes and other wild dogs on livestock enterprises. Dingoes were seen as such an immediate threat to livestock production in Australia that the dingo barrier fence, the longest fence on earth, which extends approximately 5400 km through Queensland and South Australia and stands 1.8 m tall, was built between 1880 and 1885 to exclude dingoes from preferred livestock grazing areas in the south-east.

Apart from livestock predation, dingoes have also been blamed for the extirpation of the thylacine *Thylacinus cynocephalus* and the Tasmanian devil *Sarcophilus harisii* from mainland Australia, and for stealing babies! Although that controversial murder trial in the 1980s received international recognition, dingoes continue to attract tourists to places like Fraser Island off the coast of Queensland, where nine-year-old boy Clinton Gage was killed in 2001. Dingo control programs to protect livestock enterprises, and people, are now being coordinated at federal, state and local levels of government because: a) dingoes may freely travel between states; and b) control programs have never completely fulfilled their duty to actually control some populations. In two studies, livestock predation by dingoes actually increased after control programs had been implemented and no known economic benefit to nearby livestock enterprises could be quantified by the researchers. The information presented to this point, however, says less about dingoes and more about how humans interact with dingoes.

Understanding the Australian dingo or any animal for that matter requires some on-ground experience … and a little intuition. When I began my research on dingoes in the Burragorang Valley in the Blue Mountains, one of the caretakers at the ghost town known as Yerranderie told me that he once had that instinctual feeling when walking to one of the old silver mines. So he turned and looked at the ledge behind him where a dingo was watching him walk past. Then in August 2005, I was on a short field trip with Fairfax media photographer Jon Reid and had a similar experience. We had staked out at the top dam, a spot that was frequented by two or more dingo packs, from about 1 pm until around 3 pm when three dingoes and four pups strolled past. Once they had finished drinking, playing and had walked out of the area, we tracked them back to their den and tried our hardest to sneak up on the pack but they saw us before we could get a clear photograph. Nevertheless we walked to the den, an ex-wombat burrow, and were standing on the edge when Jon said, 'I swear we're being watched'.

So I looked around and noticed that the dominant female, *Makileiko,* was sitting on an embankment less than 15 metres away watching us. Once we had made eye contact, she slowly slinked away up the hill and out of sight without so much as a whimper.

Previous to that experience, *Makileiko* was the first dingo that I had *the* fleeting glimpse of in the wild. It is known as *the* fleeting glimpse because it is about all most people get to see of a wild dingo. Her name means 'eyes' in *Gundangarra,* the local Aboriginal language, aptly so because she always appeared to be making observations when photographed by our hidden motion-sensing cameras. *Makileiko* was a perfect ginger colour with symmetrical front white paws and a black muzzle that was characteristic of 'the valley dogs' as locals called them. The dominant male in that pack was named *Gnamaiko,* meaning 'heart'. He was sable (German shepherd colour) and subsequently identified as the heart of the project, notwithstanding his apparently off-colour, he looked and acted exactly like a dingo and epitomised our study. While observations of *Makileiko* and *Gnamaiko* were made on numerous occasions either directly or through the eyes of our hidden cameras between 2004 and even the last official data collection field trip in April 2007, we never trapped them. We did, however, trap the yearling/subordinate female in their pack during April 2005. She was the first classic tan dingo-coloured animal that we trapped so I named her *Mirri,* one of the *Gundangarra* words meaning 'dingo' (see Figure 1.2). *Mirri* also was observed on the last field trip in April 2007. Unlike *Makileiko* who looked in much poorer condition then, than on previous occasions, *Mirri* was pregnant. During my project, these dingoes and others they travelled with were identified as *the dingo pack.*

The subjects of our study were the dingoes living in one of the first dingo conservation zones designated in Australia. The main aim of the study was to document as many aspects as practicable about the dingoes in the Blue Mountains so we could understand their functional role in this landscape better. To do that, we divided the aim into five main objectives:

1 to assess the genetic 'purity' of the population and to assess the genetic relatedness of individuals and packs;
2 to compare their morphometric measurements (body size and shape) and colour with those previously reported for dingoes in other studies;
3 to identify diet and seasonal dietary shift;

Figure 1.2 Two of the dingoes studied in the Burragorang Valley, *Mirri* and *Makileiko*, at the top dam during an afternoon walk in September 2005.

4   to assess changes in activity and abundance of dingoes and their prey and interactions between dietary data and activity/abundance data; and

5   to assess seasonal and short-term patterns of movement and how packs organised themselves spatially in the landscape.

Between 2005 and 2007, we trapped a total of 47 dingoes in the southern section of the Greater Blue Mountains World Heritage Area and saw many more. For one week every month for 24 months I would rake 50 sand plots and collect as many dingo scats (faeces) as possible with volunteers for data on dingo abundance/activity and their diet. We often drove more than 1000 km per field trip and sometimes hiked through dense scrub for hours or days on camping expeditions to retrieve GPS data-logging or VHF telemetry collars. Hours were spent in the NSW National Parks and Wildlife

Service helicopters and Cessna aeroplane to locate telemetry collars and retrieve GPS data. The rest of my time between 2004 and 2008 was spent reading, writing and documenting as much information as possible about dingoes and the larger species in the family Canidae. That helped form the content of this book.

2

THE DINGO IN AUSTRALIA

The family Canidae evolved some 10–12 million years ago. They are known for the possession of uniform and unspecialised dentition, with the carnassial teeth (upper-fourth premolar and lower-first molar) arranged for shearing flesh. The family Canidae evolved during successive radiations and have consistently occupied a broad spectrum of niches. These niches have included large pursuit predators, small omnivores and even herbivory. The family Canidae are one of the oldest living families in the Order Carnivora and are the only to survive since emerging in the late Eocene.

Geneticists have used 2001 base pairs of DNA sequence from mitochondrial protein coding genes and showed that all modern-day canids are found in the subfamily Caninae, of the family Canidae, under tribes known as Vulpini, or fox-like canids, and Canini, or wolf-like canids. There is debate between palaeontological, molecular and morphological research regarding the radiation of the Canini tribe around the world. There are some researchers that suggest the family Canidae began in India, while others have suggested North America. If the wolf-like canids began in North America, their arrival in Eurasia started an extensive radiation and range expansion throughout Europe, Africa and Asia (see Figure 2.1).

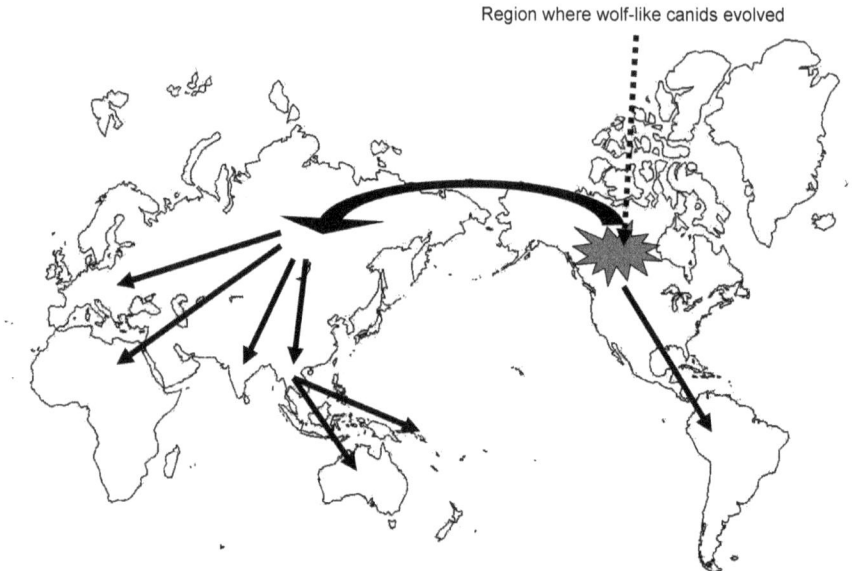

Figure 2.1 Radiation of the wolf-like canids from North America around the world. Climatic variation over time is hypothesised to be the reason so many canid forms have evolved.

Combined with the suggestion that the Canini tribe belonged to a circum-arctic fauna which survived expansions and contractions with the fluctuating global climate, the process of evolution for multiple, closely related wolf-like canids is explained. The arrival of the dingo in Australia during the late Holocene period was historic, because the family Canidae became the first globally distributed family from the Order Carnivora.

It is difficult to ascertain how many subspecies of *Canis lupus* are extant. There are many sources of information that provide estimates though none of these have provided certainty. Figure 2.2 lists 18 possible subspecies, including the controversial Eastern timber wolf and some extinct subspecies.

Phylogenetic studies cannot be satisfactorily performed on canid forms from any single continent because of their Holarctic distribution and faunal intermingling. A good example for this would be to class dingoes in Western Australia as a variety of dingo, and dingoes from the alpine regions in south-east Australia as another variety. Since that level of taxonomic classification is unnecessary for all canids, it is easier to class them based on their function in the ecosystem as hypo- (slightly), meso- (moderately) or hypercarnivorous (highly) predators. Some researchers have suggested that hypercarnivorous

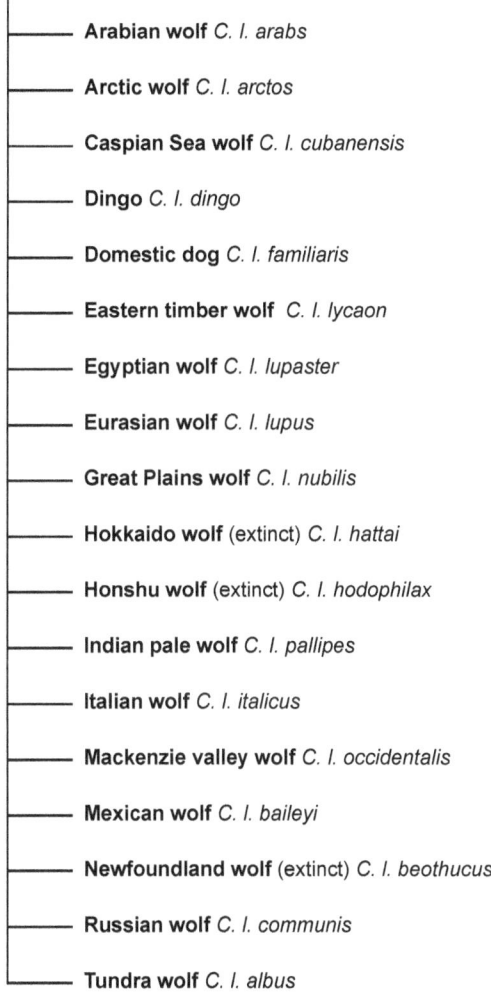

Gray wolf *Canis lupus lupus*

Arabian wolf *C. l. arabs*

Arctic wolf *C. l. arctos*

Caspian Sea wolf *C. l. cubanensis*

Dingo *C. l. dingo*

Domestic dog *C. l. familiaris*

Eastern timber wolf *C. l. lycaon*

Egyptian wolf *C. l. lupaster*

Eurasian wolf *C. l. lupus*

Great Plains wolf *C. l. nubilis*

Hokkaido wolf (extinct) *C. l. hattai*

Honshu wolf (extinct) *C. l. hodophilax*

Indian pale wolf *C. l. pallipes*

Italian wolf *C. l. italicus*

Mackenzie valley wolf *C. l. occidentalis*

Mexican wolf *C. l. baileyi*

Newfoundland wolf (extinct) *C. l. beothucus*

Russian wolf *C. l. communis*

Tundra wolf *C. l. albus*

Figure 2.2 It is difficult to ascertain how many subspecies of *Canis lupus* are extant. There are at least 18 potential subspecies of the gray wolf.

canids need to weigh more than 21 kg on average. Although some individual dingoes can weigh as little as 9 kg, such as the lone juvenile female *Mullunga* ('girl') from the Blue Mountains, their average weight is about 16 kg, and the heaviest dingo recorded so far was 21.4 kg in Western Australia. Despite the lower average weight of dingoes, my research would define them as hypercarnivorous canids because, like the African wild dog *Lycaon pictus* and the gray wolf, they have all of the other characteristics. For instance, their dentition is arranged for shearing flesh, they live communally, they hunt prey that weigh more than themselves, such as the red kangaroo *Macropus rufus* and the eastern grey kangaroo *Macropus giganteus*, and they are the largest terrestrial predator in Australia. In contrast, South-East Asian dingoes may be mesocarnivorous because they have commensal relationships with people and may compete for food resources with larger predators like the tiger *Panthera tigris*. A study on kangaroo predation by dingoes in north-western NSW also showed that dingoes had a higher predation rate on kangaroos (0.38 kg prey per kg predator per day) than wolves do on moose (0.13 kg prey per kg predator per day). The dingo can still be considered a 'primitive' canid, however, because it has retained the key characteristic of breeding once per year, similar to other primitive relatives such as the wolf, the coyote *C. latrans*, and the jackal *C. adustus*. The worldwide distribution of the dingo is currently thought to be limited to South-East Asia and Australia.

Dingoes in the Dreaming

Prior to European colonisation, indigenous Australians inhabited the continent for some 40 000–50 000 years. Dingoes were revered in some Aboriginal cultures and some sources have suggested that dingoes were used by Aboriginals for hunting. There are numerous Aboriginal cave paintings that depict dingoes across Australia including one in Nattai National Park in the southern section of the Greater Blue Mountains World Heritage Area and others in the northern section.

In 2006, Merryl Parker from the University of Tasmania explored 'the narratives within which Australians have "trapped" their dingoes', showing that Aboriginal stories of dingoes were either ennobling or were warnings to improve the human condition. Merryl wrote extensively about representations of the dingo in the Dreaming, and stated that dingo stories were grounded in empirical knowledge and myths that referred to known characteristics of the dingo. These included its regular need for water, communal family structure, territoriality and its danger toward humans. Although the dingo may have replaced the thylacine in stories, the dingo had the same positive and consistent, top-order predator role in the scheme of things. In one work, Merryl referred to Aboriginal religion as 'the grand design plan' that incorporated all living and inanimate things into one interdependent whole that permeated every aspect of daily life. Aboriginal religion explained the mysteries of birth and death and provided a charter of social and spiritual behaviour to guide each person through life. Their myths imparted the belief that heaven is here and now and that the land must be passed on to the next generation in good condition. Australian Aborigines therefore were using observations and stories about dingoes to sustain resources and society.

For fear of misrepresenting their true meanings, Dreaming stories involving dingoes won't be recounted here. The word *dingo* itself has been referred to as:

> 'a linguistic interloper with no known place in any indigenous dialect [that] pushes its way to the front to participate in any colonial discourse of negation and contempt.'

Dingo may have been the name of the animal that was pointed at when the colonists arrived and asked what that animal was called, but it is not the name used in any Dreaming stories. Merryl also showed that many stories had been misconstrued by publishers and English translators of Aboriginal dialects. In my study site, dingoes were referred to as *Binure* (old mountain dingo), *Warrigal*, *Mirrigang* and *Mirri* (all meaning dingo) and *Merrigang* (dog), while dingoes in *Pintjantjatjara* were called *Papa*, in *Alyawarra* they were called *Aringka*, and *Tiwi* people called them *Palangamwari*. *Gundangarra* dialect was used to describe and remember some of the dingoes that we encountered in the Blue Mountains. *Daoure* (earth) was trapped at the base of a big red dirt hill and as is discussed later, travelled extensively through the region, up rivers, over cliffs and along gullies. *Kurragang* (magpie) was the distinct black and white dingo that we saw for the duration of the project but could never trap. Then there was *Gimbil* (a spark), *Mikee* (lightning), *Moikke* (black cockatoo), *Burraran* (to bite), *Coy* (call to), *Mungalee* (to sit) and *Bunyal* (sun), to name a few.

Despite the history of Aborigines with the land and their religious beliefs that protected the countryside for future generations, indigenous Australians have been implicated by European Australians in causing the extinction of some species and changing landscapes through their land management practices, such as firestick farming. Alternative hypotheses provided by other researchers indicated that other environmental variables were contributing to decline in some animal populations during their time as convenors of the land. Compared with Aboriginal culture, however, the

Euro-Australian culture that exists today has permitted landscape modification to promote modern industrial economic growth and support an ever-increasing population with unknown effects on the Australian biome (the main groups of plants and animals living in areas of certain climate patterns). In stark contrast, colonial European ideology has viewed dingoes as enemies, resulting in the degradation of dingoes and as a consequence, the Australian environments they inhabit.

Current distribution

The effect of dingoes upon arrival in Australia during the first 3800–4800 years before European settlement is not well known. Based on its reproductive rate and agility, dingoes could have occupied the entire continent within 500 years. Occupation of territory and adaptation to the varied ecosystems and landscapes may have been made easier for dingoes with assistance from the Aboriginal population. Dingoes scavenging food scraps at Aboriginal camps was once reported in 1942 and it was hypothesised that domestic dogs may have originated from such interactions between hunter-gatherer societies and wild canids.

Since European colonisation and settlement of Australia, available habitat for the dingo has been severely fragmented. The dingo barrier fence, intense control programs, rapid development of towns and infrastructure

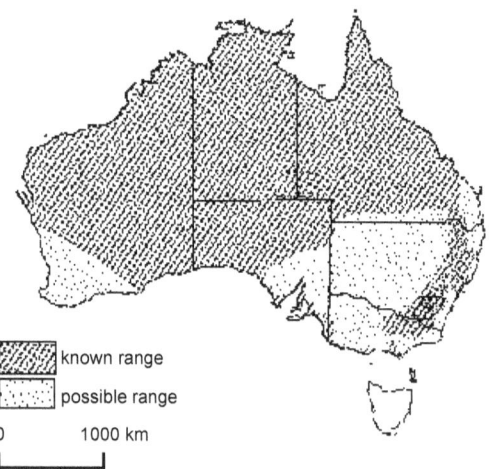

Figure 2.3 Current distribution of wild dogs (dingoes, feral or wild domestic dogs and their hybrids) in Australia. Adapted from Macdonald and Sillero-Zubiri 2004b

and land clearing for agriculture created more isolation of remaining dingo populations. Figure 2.3 presents the current distribution of dingoes and other wild dogs, including feral or wild domestic dogs and their hybrids in Australia. They can inhabit a diversity of ecosystems from tropical to temperate rainforests, deserts and all habitat types in between.

Dingoes are not found in fossil records in Tasmania so it may be proposed with some certainty that dingoes arrived in Australia after Tasmania had become an island due to rising temperatures and sea levels between 10 000 and 18 000 years ago. Dingoes inhabited all mainland Australia either in commensal relationships or as free-roaming competitors for food with Aborigines prior to European settlement and the construction of the dingo fence. As I stated earlier, dingoes have been blamed for the extinction of the thylacine and the Tasmanian devil from mainland Australia due to direct or indirect competition. The arrival of the dingo in Australia coincided with:

- expansion of the Aboriginal population;
- developments in the small tool tradition of Aborigines arising simultaneously in India;
- environmental change induced from firestick landscape management practices; and
- climate change.

The dingo may therefore assume a shared responsibility with Aborigines and climate change for the mainland extinction of the two largest endemic marsupial carnivores. This indicates that despite popular opinions in which dingoes are suspected of having a negative impact on Australia prior to European settlement, other explanations are equally as plausible. Further evidence of such a theory is the prevalence of most small-medium marsupials such as bandicoots and the bilby *Macrotis lagotis*, which are now threatened with extinction due to European settlement and the introduction of foxes and cats. Expansion of European settlement, developments in agriculture and landscape changes resulting from agriculture since European colonisation can, similarly, assume responsibility for the persistence of dingoes despite eradication efforts. Water abundance has, for example, increased prey availability and potentially assisted survival of some populations through droughts. Prevalence of domestic dogs with settlers has also provided new opportunities for dingoes to mate, posing a threat to the survival of 'pure' dingoes due to hybridisation.

Approximately 95 years after the dingo fence was erected (or 190 years since colonisation), the first major research on the dingo was published. A general overview on the biology of the dingo including genetics, growth and ageing, reproduction, behaviour and physiology was first available in 1977. Soon afterwards more details on social biology, movements and the effects of control efforts in Western Australia, central Australia and coastal New South Wales and Victoria were reported at the Australian vertebrate pest control conference in 1978. Findings from research on the relationship between dingo skull allometry (size and shape) as a descriptor of 'purity' were first published in the early 1980s.

The threat of hybridisation to 'pure' dingoes with domestic dogs has been listed as a key threatening process to their survival on the International Union for the Conservation of Natural Resources (IUCN) Red List of Threatened Species. This listing was elevated from lower risk in 1996 to vulnerable in 2004. As a result of hybridisation with domestic dogs, 'pure' dingoes are expected to be extinct in the wild within 50 years. The problem, however, is not that the dingo may be going extinct, nor that the dingo remains relatively unprotected by Australian legislation; the problem is how to manage the hybridisation process. In March 2000, land management authorities in NSW designated 209 reserves, from more than 1269 available reserves, as dingo conservation habitat. These were the first natural areas recognised for their importance to dingoes in Australia, and showed that the culture of dingo management was shifting towards conservation.

Other researchers have recently published their studies that showed dingoes may limit the abundance of foxes and cats and assist in the conservation of native Australian marsupials. Some data also has shown that dingoes can't be entirely blamed for the extinction of the thylacine and the Tasmanian devil from the mainland. For instance, the mechanical loading (bite pressure) from a thylacine is much stronger than the pressure from a dingo. In addition, thylacines were heavier than dingoes on average and there are historical reports that, like dingoes, thylacines also hunted and moved in groups. Other researchers have showed that dingoes have positive effects for native marsupials by suppressing mesocarnivorous mammals such as the red fox *Vulpes vulpes* and the domestic cat *Felis catus*. Some of the researchers used the dingo barrier fence to show, for instance, that foxes were more abundant in areas that dingoes were absent, and native marsupials were more abundant when dingoes were present.

3

WHAT IS A DINGO AND HOW DOES IT DIFFER FROM A DOMESTIC DOG?

For the last 220-odd years, humans have pondered the origins of the dingo. John White, Surgeon-General to the First Fleet and settlement at Port Jackson titled the first illustration of a dingo in 1789 'dog of New South Wales' (see Figure 3.1), and stated the following in his journal:

> 'This animal is a variety of the dog, and, like the shepherd's dog in most countries, approaches near to the original of the species, which is the wolf, but is not so large, and does not stand so high on its legs.

> The ears are short, and erect, the tail rather bushy; the hair, which is of a reddish-dun colour, is long and thick, but strait. It is capable of barking, although not so readily as the European dogs; is very ill-natured and vicious, and snarls, howls, and moans, like dogs in common.

> Whether this is the only dog in New South Wales, and whether they have it in a wild state, is not mentioned; but I should be inclined to believe they had no other; in which case it will constitute the wolf of that country; and that which is domesticated is only the wild dog tamed, without having yet produced a variety, as in some parts of America.'

Figure 3.1 Surgeon-General John White's 'Dog of New South Wales', published in 1789.

Nearly 200 years later in 1975, Neil Macintosh published research titled *The origin of the dingo: an enigma* which classed it as puzzling, ambiguous and inexplicable. We now know through numerous means that dingoes originated in South-East Asia, potentially the same birthplace of the domestic dog. In 2002 results published from a study showed that there was a high possibility that all domestic dogs came from one single gene pool in East Asia approximately 15 000 years ago. That means that dingoes had possibly been interbreeding with domestic dogs for 10 000 years before their arrival in Australia. In 2004, a study by a similar research group showed that all dingoes may have originated from one single pregnant female. Reproductive isolation for this female and her pups would have been the start of Australia's genetically unique dingo population if other dingoes were not traded with Aborigines or introduced to Australia ever again. Scientists have shown that some dingoes were reintroduced to Asia based on the distribution of lice and there were multiple trade routes for people between Asia and Australia during the last 10 000 years. Even in Captain Cook's *Journal During the First Voyage Round the World* there is evidence that dogs and hogs were freely traded between travellers and people for food, companionship and goodwill. Any additional introductions of dingoes since that first female would have started the process known as hybridisation. The old trade routes connected Australia with Thailand, China, Borneo, Papua New Guinea and Indonesia.

Introgression of domestic dog genes with dingo genes has been identified as a conservation dilemma for the preservation of 'pure' dingoes. Attempts to define purity of wild dingoes have been limited to comparison of morphological measurements, skull allometry and genetic biotechnology. Scientists originally cross bred dingoes with domestic dogs in captivity over six years to investigate the possibility of hybridisation in the wild. Results from those studies have since been used to classify wild living dingoes as hybrids or pure. Some researchers have questioned the validity of this as a technique and concluded that what appear to be hybrid dingoes on first impression had more dingo traits than domestic dog traits. One trait, for instance, comes from the core of biology, the one annual reproductive cycle of primitive canids.

Contemporary studies on the identification, biology, behaviour and ecology of Australian dingoes were published from the 1970s onwards. Dingo population numbers were severely reduced on the south-eastern side of the dingo fence at this time, which limited the distribution of dingoes to mountainous, forested habitats. Topographic and geographic isolation of dingo populations around Australia has often been overlooked, largely because studies have been comparing skull characteristics from wild populations with skull characteristics from the research on hybridisation between dingoes and dogs in captivity. In a similar sense, isolated, captive-bred dingo populations that were subjected to controlled breeding programs were tested to identify 'pure' dingo genes, and used to discriminate pure dingoes from hybrid dingoes.

Genetics

In 1975, researchers reported that there were no discernible differences between blood proteins of dingoes compared with domestic dogs. With the development of genetic biotechnology techniques, however, geneticist Alan Wilton from the University of New South Wales compared microsatellite variation in Australian captive dingoes with domestic dogs in the late 1990s. Microsatellites are short sequences of DNA that are repeated consecutively, and the number of times that the sequence is repeated varies between individuals, within populations and between species. Less variation between individuals indicates a higher degree of relatedness. Fourteen previously described microsatellites from domestic dogs were compared with 16 captive dingoes and 16 domestic dogs. When compared, results showed that there were differences between microsatellite variation between captive dingoes and domestic dogs. The level of variation in the captive dingo samples used was distinctly lower than the level of variation observed in domestic dogs.

In 2001, Alan Wilton applied his dingo purity tests and compared 77 captive dingoes with 55 mixed breed domestic dogs and 50 'wild dogs' from one site in south-east Australia. He found that the majority of microsatellites tested showed similar distributions within their genetic group. This meant that the domestic dogs were genetically similar to other domestic dogs, wild dogs were genetically similar to other wild dogs and captive dingoes were genetically similar to other captive dingoes, an outcome that would be expected when using a test suited to show genetic relatedness between individuals within populations. Once again, variability was lower in the captive dingoes than in the dogs, showing that dingo breeders had maintained purity of line in their captive populations. In addition, it was not specified whether the wild dogs that Wilton sampled were grouped with domestic dogs as 'dogs' when genetic variability was discussed.

Due to that study, a technique to ascertain a calculated probability that a genetic sample was more likely to be from a pure dingo, a hybrid or a domestic dog based on microsatellites was developed. Although no definitive answer on purity or hybridisation could be attained, animals could be assigned to categories of dingo, three-quarters dingo, hybrid or domestic dog. As the number of tests increased the more likely it became to detect past dilution events, and to determine differences between populations of dingoes across Australia. Using this information, Alan Wilton implied that the conservation value of a population could be determined and screened through genetic means.

To estimate the time of the first introduction of dingoes to Australia, scientists investigated the distribution of mitochondrial DNA (mtDNA) from 211 dingoes, 22 dogs and 19 pre-European archaeological dog samples in 2004. These samples were compared with 654 dog and 38 wolf samples from another study that tested the origin of domestic dogs in 2002. The sequence variation of mtDNA in dingoes was very restricted when compared with dogs and wolves. Their results indicated that all dingo mtDNA types originated from one type, type A29, because it was observed in 53% of samples. Based on an estimated wolf-coyote divergence of two million years ago, the arrival of the dingo in Australia was calculated to be between 4600–5400 years ago. A considerable difference for the mean distance to A29 was also observed between dingoes in the north-west of Western Australia and other parts of Australia. This discrepancy was treated as local genetic drift. Observations of A29 in samples from East Asia, New Guinea and Arctic America showed that 47% of genetic samples were unaccounted for and the discrepancy in Western Australia implied that the introduction of dingoes to Australia may have occurred more than once. The researchers refuted this finding because a number of other mtDNA types, not observed in Australian dingoes, were observed in samples from islands surrounding Australia.

Another important aspect of dingo genetics involves phenotypic differences between geographically isolated groups but these are yet to be addressed in great detail. Researchers in Western Australia suggested that localised gene pools must have existed where black and tan colouration and white colouration occurred at higher frequencies in some packs more than others. This is similar to variation of colouration in rosellas, where the birds found in the top end of Australia have black hoods, in south-east Australia they have red hoods and in north-east Queensland they have yellow hoods. Observations such as this on a range of geographically isolated populations of dingoes across Australia could become increasingly important for conservation efforts to manage dingo populations effectively.

Biological variation in the family Canidae

Dingo skulls have characteristics for functions linked with predation and when compared with domestic dog skulls, dingoes have a longer muzzle, a flatter cranium and a higher nuchal crest (a tuft, ridge or projection on the head). Main teeth in dingoes are larger and the canine teeth are longer and slightly more slender for mastication and shearing flesh: characteristics most often used to identify top-order predators.

Victorian scientist, Evan Jones, first questioned the validity of classifying dingoes as pure or hybrids using calculations of skull allometry in 1990. He stated that the wild canids in Victoria's eastern highlands could not be called domestic dogs but could not be called dingoes either because some measurements recorded were outside of those specified in past research.

In conclusion he stated that:

- there were no reliable methods available to classify the present canids into categories of purity;
- canids with coat colour, canonical score and physical conformation of 'pure' dingoes were present;
- less dingo-like canids could not be called hybrids because genetic structure could not be accurately quantified;
- extreme, domestic dog-like traits for head, body and limb shapes were not present;
- a uniform, dingo-like physique was common;
- reproductive patterns did not follow patterns expected for feral dogs (biannual breeding cycles); and
- feral dogs and/or hybrids were conclusively few in number.

Researchers who studied dingo skulls sampled in Queensland also were equivocal about why, where and when hybrid or pure specimens were collected. Evan Jones had observed an increased range of variability in skull shape, coat colour and physique in Victoria compared with dingoes that were considered to be 'pure' from central Australia. Based on skull scores he suggested that introgression had occurred but could only speculate an explanation for hybridisation. Comparative analysis from 320 dingoes sampled in the Victorian highlands before 1974 with dingoes sampled in his study during the 1980s showed ginger/tan coloured dingoes had decreased by 9% and black and tan animals had increased by 9%. Evan did not record sable colouration although it was recorded before 1974 in the Victorian highlands and had been recorded in central Australia. Brindle colour had decreased by 2%, black colour had increased by 6%, straight white, another 'pure' dingo colour, decreased by 4% and patchy colouration was not recorded.

Contemporary studies on inheritance of coat colour, skull morphology and other phenotypic descriptors for vertebrates from around the world have shown that use of these techniques alone to characterise populations may be limited. For instance, skull allometry can vary due to a range of environmental circumstances such as mechanical loading. Mechanical loading is the compressive, tensile or shearing force applied to a solid object during a bite. In terms of dingo ecology, mechanical loading is the pressure of exertion used by dingo facial muscles to shear flesh and masticate prey. Mechanical loading can, therefore, alter the skull shape of dingoes as an epigenetic process.

Domesticating the silver fox

In an attempt to answer questions regarding the evolution of the domestic dog, researchers in Russia selectively outbred silver foxes *Vulpes vulpes* from fur-farm bred populations. The general aim of the experiment was to observe adaptations in behavioural traits from foxes that retained standard phenotype, biological function and wildness to that of domestication and conditioning of fox behaviour due to the presence of human beings. From 50 000 offspring studied over 40 years, they showed that adaptation responses to domestication had occurred within six generations and phenotypic variations were commonly seen due to experimental selection pressure.

Phenotypic variations in captive-bred foxes included colour changes such as yellowing or brown mottling and piebaldness, floppy ears, curly and short tails, abnormal tooth-bite (under-bite, similar to bulldog breeds) and other craniological changes (see Figure 3.2). There was also a change in seasonal reproduction patterns observed, whereby oestrous cycled in some vixens during autumn and in spring, though extra-seasonal matings were rare.

Changes in dimensions of the domesticated silver fox skull included a widening of the facial skull and a decrease in the width and height of the cerebral skull. These changes were more prominent in males selected for domestication, whose skulls were smaller in almost all proportions when compared with male foxes bred for fur.

Figure 3.2 A piebald domesticated silver fox (top left); a curly tailed domesticated silver fox (top right); a mottled domesticated silver fox (bottom left) and a mottled domesticated dog (inset); a comparison of non-domesticated silver fox farm silver fox skull (left) with a domesticated silver fox skull (right) showing variations in width and length (bottom right).
All photos: Lyudmila Trut

Dingo skull allometry

If we compare the cranial morphology of wild and domestic dogs, the skull length within all dog breeds mimics exact allometric dwarfism due to a lack of developmental variation when they are compared with wolf-like canids. One exception was that gray wolves tended to have longer teeth than similar sized domestic dogs – a difference that was also observed between dingoes and domestic dogs. Figure 3.3 illustrates the susceptibility of the neonate canid skull to developmental processes. Further, alterations in the timing and rate of postnatal growth in the canid skull can assist the evolution of new cranial proportions.

Observations made in the silver fox study indicated that within isolated populations of canids, major genetic and phenotypic changes can occur within very few generations. Dingoes from south-east Australia may therefore differ morphologically to dingoes from central and northern Australia due to the construction of the dingo fence in 1885 and other changes in land use that have isolated or fragmented dingo populations. It has been speculated that the areas where dingoes were less controlled, in the north-western half of Australia, comprised mostly 'pure' dingoes according to canonical analyses of skull measurements (see Figure 3.4). Such speculations are consistent with hypotheses of epigenetic effects on skull allometry. If control programs were to change social and feeding habits, then skull allometry may also change in response.

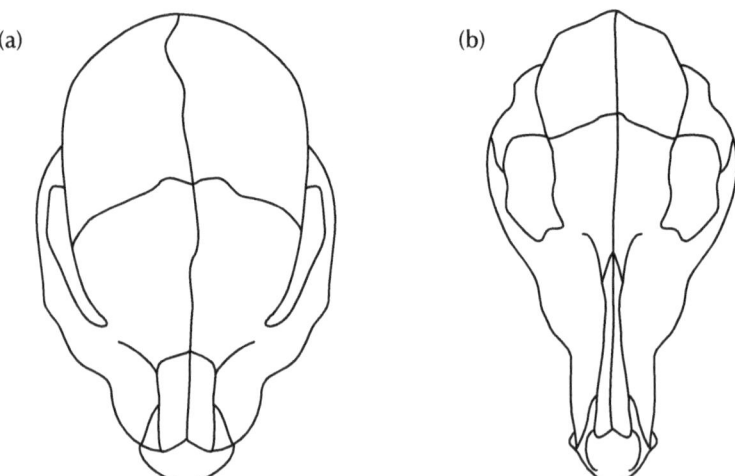

Figure 3.3 Comparison of neonate canid skull (a) with an adult canid skull (b) shows the profound developmental alterations that occur from neonate dog to adult dog.

Source: After Wayne 1986.

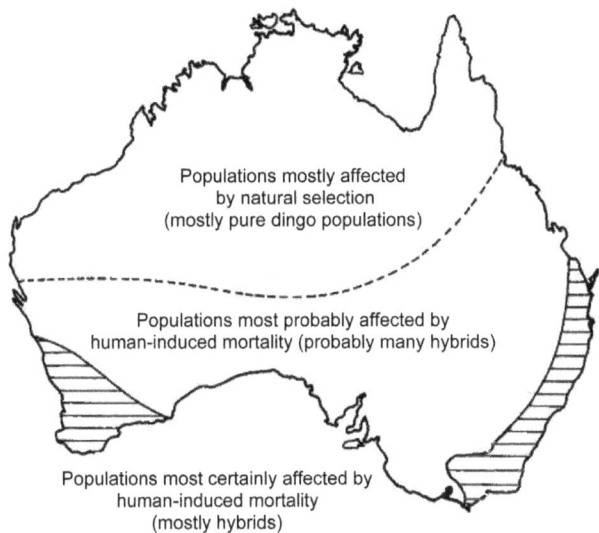

Figure 3.4 Locations of dingoes showing 'hybrid' canonical scores due to selection pressure from being persecuted and locations of dingoes showing 'pure' canonical scores due to natural selection pressure. Adapted from Corbett (2001). The words in brackets are the original descriptors of the dingo populations used by Corbett.

Three out of eight parental dingoes in the original sample of 'pure' dingoes from central Australia used to identify what a pure dingo was returned 'intermediate' canonical scores (see Figure 3.4). In that study, 'intermediate' appears to be a convenient misuse of the word 'hybrid'. Cranial changes were recorded within six generations of foxes reared in experimental conditions. Studies in the 1960s also reported cranial changes in captive-bred wolves. It is evident from these studies that life in captivity from birth can affect cranial development in the family Canidae. It may also be proposed that food supply and methods of hunting could create rapid adaptive morphological response to environmental conditions. Observations made during past experiments of captive-bred, primitive canid species may, therefore, have been affected by the food the subjects were fed.

Epigenetics

Epigenetics is a broad term that means *beyond genetics*. Phenotypic variations, such as changes to skull shape due to the pressure exerted by facial muscles when biting, are beyond genetics because they are not caused by DNA sequence change. Such variations can be both heritable and non-heritable traits. Natural selection can produce most of these changes provided that there is genetic variation on which to work; however, epigenetic processes can induce phenotypic variation without requiring nuclear DNA variation. Some authors have used slightly different interpretations of epigenesis, the theory of epigenetics, but in general it is used to describe phenotypic variations that are not explained by conventional Mendelian genetics.

Various chemicals that are epigenetic 'marks' sit atop genes and provide the genes with basic instructions such as to turn on, or to turn off. The activation or inactivation of genes is generally driven by environmental factors like diet, stress and parental nutrition. Epigenetic marks have been demonstrated in a range of vertebrate and invertebrate species and biological differences have occurred within species in one generation. Environmental processes can influence DNA from as early as the development of gametes, which doesn't occur in male humans approximately until age 11.

How does this apply to dingoes? First of all, dingo researchers once detected a major difference in reproductive response by dingoes between flush and drought periods in arid central Australia. Males also showed marked seasonal variation in testis size and weight during the annual breeding season which suggests that epigenetic marks may vary in dingoes in every breeding season. Reproductive suppression due to social constraints is another factor (stress) that may affect dingo epigenetics, but a lack of food due to drought has been suggested as having a greater influence on sperm production. Observations of epigenetic effects in humans also have been linked to times of famine.

Secondly, dingoes are a top-order predator and may in one instance suffer from bioaccumulation of chemicals like endocrine-disrupting compounds (EDC) or pesticides and herbicides. These chemicals can be long lasting in the environment and may be consumed by the herbivores that dingoes predate. EDC are external substances that act like hormones in the endocrine system and disrupt the function of internal hormones. This is similar to the effects that DDT had on raptor populations where the eggs laid had thin, weak shells that were easily crushed by the parents when they landed on their nest. EDC are known to be particularly persistent in long-lived organisms feeding at the top end of the food chain. They can be naturally occurring or synthetic and have been shown to either activate or inhibit endocrine pathways, depending on the receptors they bind to during fertilisation and embryonic development. Variation in coat colouration is one area that may be affected by epigenetic marks, and this has been observed in laboratory-reared mice.

Studies have shown that EDC can contaminate water resources and their effect on ecosystems is of concern. Industrial processes (such as wool scour, livestock husbandry and agriculture) coupled with large (residential) and small-scale sewage treatment plants are potential sources of EDC throughout Australia. Studies of the impacts of EDC have shown that mammals can be affected through decreased fertility, demasculinisation and feminisation, and alteration of immune function.

Endocrine-disrupting compounds and other epigenetic activities are applicable to dingo ecology for two reasons: 1) dingoes are found at the top end of the food chain; and 2) research has shown lactating dingoes consume the urine and faeces excreted by their young. Therefore if a lactating female has been exposed to EDC or other chemicals, they will be recycled and become concentrated. The quality and quantity of milk lactated by a female dingo to newborn pups can therefore also alter the gene expression in dingoes. Any genetic models to define a species like the dingo now need to consider:

- direct genetic effects from the genotype of the progeny;
- heritable epigenetic affects from the genotype of the progeny;
- indirect epigenetic effects from interactions of two separate genotypes; and
- non-heritable environmental effects, such as EDC and bite pressure.

Skull development for young dingoes may be dramatically affected by the food that they eat and the pressure exerted by their facial muscles while they're eating. This mechanical loading will obviously vary per dingo population for the local prey and other food types available. Due to processes that are beyond genetics, characteristics originally used to define a pure dingo from skull size and shape suddenly become subjective, as does the use of genetic techniques. Due to temporal changes and the range and changing nature of human influences across ecosystems, epigenetic effects on dingoes are likely to vary across their range. Differences in genetic variation, skull allometry and colouration between populations are perhaps, therefore, not related to hybridisation alone (see Figure 3.5).

As discussed in Chapter 1, pure dingoes are reported to have three general coat colours: tan, black and tan and white. 'Pure' dingoes may also be sable, brindle, black or even black and white depending on how historical reports were interpreted. The fox-farm experiment showed that coat colour

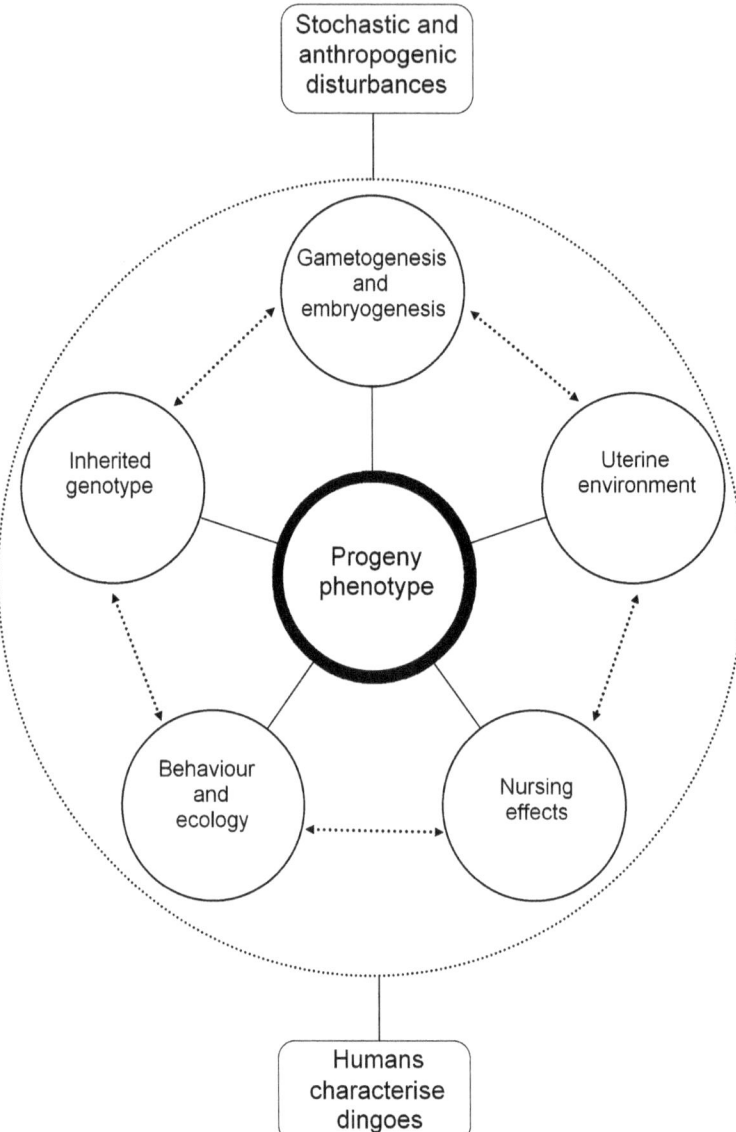

Figure 3.5 A general causal model describing dingo development (adapted from Atchley *et al.* 1991). Natural and anthropogenic disturbances (broken circle) affect all of the primary causal effects (solid circles) that interact (broken arrows) to create the phenotype of the individual (thick circle). The magnitude of natural and anthropogenic disturbances or other dissimilarities (such as available food items) at any particular geographic location will cause variations in dingo phenotype and other characteristics that humans use to define dingoes.

can vary from a uniform silver coat to a range of unexpected colours within a selectively bred population. Thus the question arises: if a high-density, outbreeding population of wild dingoes were living in optimal wild conditions, similar to farm-bred, domesticated foxes, would selection pressure cause variations in phenotype? Results from the fox research indicated substantial variations could occur within six generations, the same length of time that the original dingo-dog hybridisation study endured.

What defines a 'pure' dingo and are there any left?

It is all too easy for media, and land and wildlife managers to use words like *mongrels, wild dogs, dingo hybrids* and even the word *dingo* to alter public perceptions about dingoes or to alter policy. Regional affairs reporter for the *Sydney Morning Herald*, Debra Jopson, titled her report 'Farmers pay high price in country gone to the dogs' on 26 October 2009. Debra reported that a local farmer:

> 'reckons the feral dogs are at their worst in his 72 years, breeding in Sydney Water's catchment and national parks around the Wollondilly River and travelling to nearby farms to kill sheep and calves.'

Eight months earlier, in the same newspaper, science editor Deborah Smith titled her article, 'Dingo behaviour gives locals paws for thought'. She reported results from my study on the dingoes in Sydney Water's catchment and national parks around the Wollondilly River and quoted me:

> 'Importantly there were minimal, or no visits to farmland by the GPS-tracked dingoes.'

These are two polar-opposite views about the same population of dingoes. To some people this population consists of nothing more than mongrel feral dogs, while to others it consists of hybrids, dingoes or both! To resolve this conflict, we need to look way back in time to find out what a dingo is and the only way to do that is through genetics.

Research on genetic variation in the family Canidae has embraced colour, population structure, origin and genetic diversity. In this section the importance of identifying population structure in communal-living, wild canid populations is presented. Perceptions of purity and the success or otherwise of microsatellites to define dingo purity are addressed, as is the subsequent limitation of research on mtDNA to identify an origin for dingoes. Research on the genetic relatedness of the dingo to other canid species is used to illustrate the position of the dingo genotype in the phylogenetic tree of the family Canidae.

Figure 3.6 shows the divergence of a number of domestic dog breeds from the wolf and coyote genetic sequences. The dingo appears to be a more

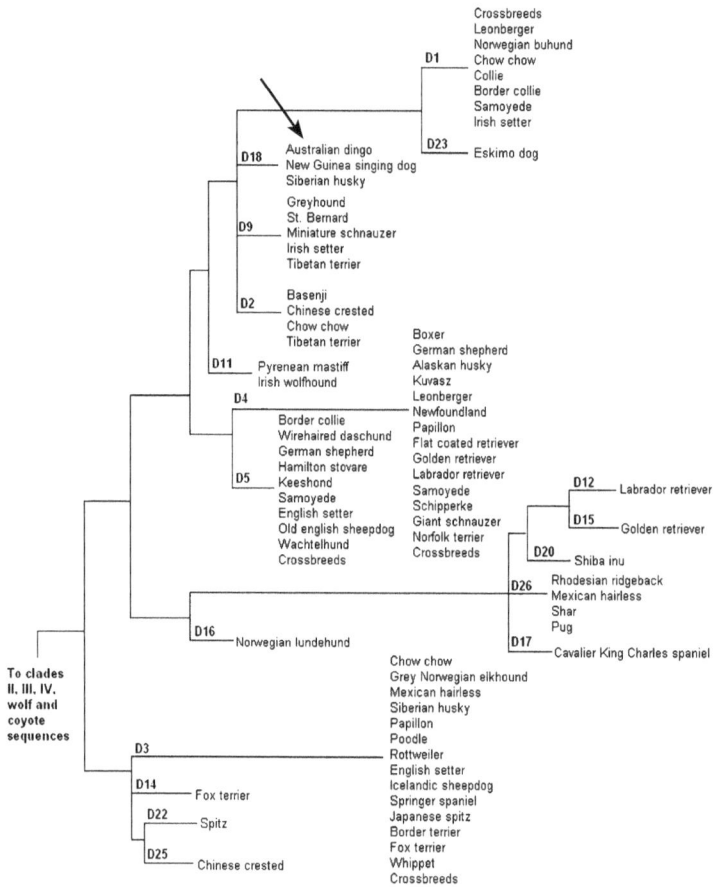

Figure 3.6 Expanded view of neighbour-joining tree of domestic dog mtDNA haplotypes within clade I (of IV) based on 261 base pairs (bp) of control region dog sequences (adapted from Vilà *et al.* 1997) to illustrate the apparent position of the dingo (black arrow) among domestic dog breeds and how genetic data can provide variations for interpretation (N.B. multiple haplotype lineages for the Leonberger, chow chow, border collie, Irish setter, Tibetan terrier, German shepherd, English setter, samoyede, golden retriever, Labrador retriever, Siberian husky and crossbreeds).

recent addition to the world of dogs. Figure 3.7 is an alternative for divergence of domestic dog breeds and implies the opposite. That is, the dingo diverged as a separate breed earlier than domestic dog breeds and that the dingo may have been the last subspecies of the wolf before the origin of domestic dogs.

So what is a dingo? Ignoring purity altogether, Figure 3.7 suggests the dingo separated from the evolutionary line from wolves and coyotes before most domestic dog breeds. Longer mtDNA sequences were used to construct Figure 3.7, which provides a more reliable measure of relatedness. Figure 3.6 shows dingoes arising after the separation of some domestic dog breeds. Other studies have reached similar conclusions and deduced that 80% of domestic dog mtDNA haplotypes, including the dingo and other ancient breeds, fall on Clade I, the most diverse clade of dog sequences. Originating in 'the New World' and Oceania, ancient dogs were indistinguishable from

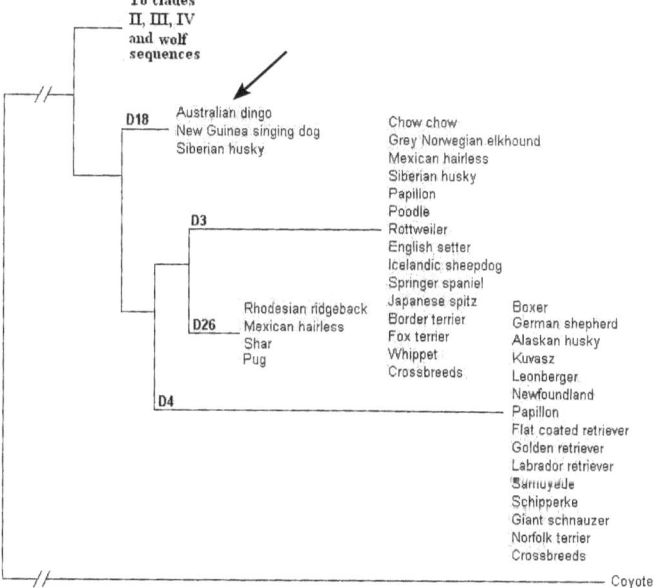

Figure 3.7 Expanded view of clade I (of IV) of neighbour-joining tree of eight wolf and 15 dog genotypes within clade I (of IV) based on 1030 base pairs (bp) of control region dog sequences (adapted from Vilà *et al.* 1997) to illustrate the apparent position of the dingo (black arrow) among domestic dog breeds (N.B. two haplotype lineages were found for the papillon breed). Labels for identification of similarities or dissimilarities between 261 bp and 1030 bp sequences and indications for bootstrap support have been excluded.

Eurasian domestic dogs, a conclusion that is consistent with Figure 3.7. Scientists have claimed that the dingoes of Australia formed as a result of inbreeding due to low population numbers within the last trickle of domestic dogs entering Australia from Asia. As stated earlier, however, there is potential that it was not a trickle of dingoes between Asia and Australia. Introgression from dingoes of Asian descent may have been common for the 3000–4000 years before Europeans arrived and may be considered a form of hybridisation, casting further doubt over the *concept* of genetic purity. Presumably most genetic studies sampled captive dingoes but no research on dingo purity was published until 1999, two years after the research that showed dingoes also group with domestic dogs on the most diverse clade of dog sequences. There is an imperfect record of dingoes in captivity and there are no records to show when the captive dingoes used for reference genetic sampling became isolated from wild dingoes. The notion of dingo 'purity' therefore becomes a construct of human thought rather than the end result of an evolutionary lineage.

Some scientists have attempted to define the phylogenic relationships of all canid species. Data from protein coding genes and morphological, developmental, ecological, behavioural and other characters was obtained and used to demonstrate that the various phylogenies of the wolf-like canids show little agreement. Instead, all that was identified is a spectrum of *possible* canid clades, or multiple origins of domestic dogs. But what about the dingo? Dingoes have been clustered with the New Guinea singing dog *C. l. hallstromi*, the African basenji, the greyhound and Siberian husky without explanation. Close clustering of the dingo with the New Guinea singing dog is understandable due to the close proximity of Papua New Guinea to Australia. Greyhounds have been assigned to a cluster of European herding dogs that then branched into European hunting dog breeds and the African basenji clustered separately to the European greyhound. The Siberian husky has clustered within the nine breeds most closely related to the wolf, which is also consistent with the dingo. But the dingo remains an enigma. In 2010, results from a study using Single Nucleotide Polymorphisms, a new genetic analysis providing broader genome coverage and higher quality data than can either microsatellites or mtDNA, were published and showed similar results to above. Dingoes are one of the closest living relatives of the gray wolf and, as the above suggests, they are most likely the last form of primitive canid before the domestication of the dog.

Dingo genes

Whenever geneticist Alan Wilton characterised dingoes based on their relatedness to captive dingoes he had to combine estimated likelihoods that the samples were from pure captive dingoes or from domestic dogs. The likelihoods were then applied to an algorithm to assess whether the individual alleles originated from a domestic dog population, a captive dingo population, a hybrid population (50% dog, 50% dingo) or a three-quarter (3Q) dingo × domestic dog hybrid population (75% dingo, 25% domestic dog). Average scores from 3Q dingoes in reference to captive dingo groups were greater than 0.58 so any samples from pure dingoes are expected to have values higher than 0.58, although 0.5 is considered a reasonable medium. Some alleles were never, or very rarely, observed in reference captive dingo samples. Genetic types that were at least 10 times more common in dogs than dingoes were defined as dog-like alleles. The presence of dog-like alleles in a sample is considered indicative of domestic dog ancestry within the captive-bred lineage.

Using these data, individual animals in a population can be designated as 'pure' dingoes, dingo × domestic dog hybrids or domestic dogs. Average frequencies of alleles shared between captive dingoes and domestic dogs are then used to broadly classify animals as having 25%, 35% or 50% domestic dog genes and samples can be assigned to one of seven categories outlined in Table 3.1.

One aspect of the methods used to characterise genetic purity was that the effects of comparing captive-bred tame dingoes with wild-born dingoes were not accounted for; they were totally absent. Earlier I showed that

Table 3.1 Assigned scores of dingo purity or percentage of hybridisation.

Score	Description
1	Pure dingo – high average 3Q, no dog-like alleles
2	Likely dingo – acceptable average 3Q in reference range
3	Likely hybrid with small amount of dog – average 3Q just below reference range >75% dingo
4	Less than 75% dingo – average 3Q less than 0
5	Less than 65% dingo – average 3Q less than –0. 1
6	Less than 50% dingo – average 3Q less than –0.25
7	No detectable dingo ancestry – average 3Q less than –0.5

characteristics such as coat colour, skull allometry and morphology can be affected by a number of factors that are beyond genetics and which change the characteristics of animals in captivity and in the wild. This effect has ultimately caused selection bias in sampling for scientific experiments because many researchers chose to sample animals that 'looked like dingoes'. It may alter perceptions about the dingo as a breed. Alan once told me that the dingoes on Fraser Island carried alleles that were not found in mainland Australia dingo populations. Data are yet to be published, but genetic variation such as this suggests captive-bred dingoes also may not represent the full spectrum of dingo alleles.

In the mtDNA study, more than half of the samples (141 of 211) for wild and captive dingoes were from south-east Australia. The dingo mtDNA data collected were compared with pre-European archaeological dog samples from Polynesia, 654 domestic dog samples and 38 wolf samples with anonymous life histories to track the trail of the dingo from South-East Asia to Australia. Although examination of specimens from museums and archaeological deposits is needed to provide a better baseline for assessment of dingo purity, even they may not be representative of dingoes Australia-wide. This is similar to the microsatellite variation observed between captive dingoes and Fraser Island dingoes.

The situation at hand then is that dingo purity has been defined using: a) topographically isolated, and in most instances caged dingoes; and b) an mtDNA haplotype specific to captive and wild dingoes with unspecified levels of genetic purity. Savolainen *et al.*, however, made the following statement in their study of dingo mtDNA:

> 'Today there is a large amount of hybridization between dingoes and domestic dogs in eastern Australia, but for this study the wild dingoes were sampled based on similarity in appearance to dingoes to exclude as far as possible feral dogs and dog-dingo hybrids, and the absence among the dingoes of any mtDNA types found in European dog breeds indicates a low degree of hybrids among the sampled dingoes. However, because of the maternal mode of inheritance of mtDNA, hybridization between male dogs and female dingoes would not be identified by these mtDNA analyses. It can be noted that, similar to dingoes, the New Guinean dogs had the mtDNA type A29 and a unique type differing from A29 by one substitution. These so-called New Guinea singing dogs are feral and show some morphological and

behavioral similarities to dingoes. A common origin and some gene flow between the two populations is therefore possible.'

Claiming that hybridisation in eastern Australia is extensive without reference or evidence of this topic earlier in the mtDNA research article is speculative. Selecting samples from wild dingoes based on similarity in appearance to dingoes adds to the speculation in the interpretation of results and in identifying dingoes. Microsatellite markers in addition are best used to *infer* relatedness. That means that without further study on purity and relatedness of wild dingo populations, each pure dingo only appears to be pure because it has markers that are common to that captive-bred lineage. Purity tests instead have shown genetic relatedness of the captive dingo population, not the purity of dingoes across Australia. Maybe then, future genetic research on dingoes should involve sampling DNA from widely distributed dingo populations, including captive-bred groups, and assessing genes for epigenetic marks. Observations could then be made with regards to how extensive genetic variation is between all of the different populations found in the north, south, east and west of this vast continent, hence validating estimates of dingo purity. In a study of that calibre, it would be possible to identify the genetic relationships between individual dingoes, and spatial relationships of dingo packs in the wild.

Genetic structure of wild dingo populations

As alluded to previously, microsatellites serve as a useful tool to infer relatedness between individuals within populations. Dingo researcher Peter Thomson once used observed variations in coat colours between packs to speculate about the genetic population structure of wild dingoes in Western Australia. In his study area, Peter noticed that there was an unusually high frequency of black and tan and white dingoes in one pack, compared with other packs. He used this as evidence of the close genetic relationships that existed within packs and suggested that pack members were reproductively isolated from adjacent packs.

Results from analyses of microsatellites and mtDNA used in a study to investigate genetic relationships between wolf packs were expected to show that the packs consisted of a dominant breeding pair and their offspring. However, nine of 27 packs sampled showed that one pair and their offspring did not account for all the individuals in a pack. Although each pack comprised closely related individuals, dispersal of near relatives among packs or the packs that shared a common boundary was common and contributed to population structure. These results showed that genetic relatedness data was capable of identifying packs that were derivatives from larger packs. Similar research on African wild dogs also found that the dominant pair and adults within each pack were unrelated. The high level of unrelated individuals within packs, combined with dispersal behaviours, was speculated to limit opportunities for incestuous mating.

Genetic research like these studies and dingo purity studies, however, remains controversial. One criticism of genetic research on the family Canidae stated that the use of mtDNA to assess genetic variability within a population was the prevalence of biased genetic expression due to the maternal inheritance of mtDNA. Microsatellite DNA was instead suggested to be best used to identify familial relationships because of its high mutation rate and maternal and paternal inheritance. Since dingoes are hypercarnivorous, communal-living canids, a similar genetic structure to that reported for wolves and African wild dogs can be expected for dingoes.

We put this to the test using DNA samples from dingoes in the Blue Mountains. Microsatellites from 53 wild dingo samples were compared with each other to infer relatedness. Analyses of genetic structure firstly identified two genetic groups consisting of highly related individuals with minor overlap of genes. One of the genetic groups split into six or seven familial group clusters. When we illustrated their level of relatedness with their capture location on a map, it became clear that the related individuals generally were trapped in discrete topographic areas. Few individuals that were trapped outside of these areas, mainly males, had travelled extraterritorially during the breeding seasons in 2005, 2006 and 2007.

Of the animals sampled, 81.1% of the dingoes shared more than 75% of their genes with other dingoes from their genetic group. In comparison with other studies, research on domestic dog breeds also has shown that different breeds were genetically distinct – much like the results from tests of genetic purity in captive dingoes. Each genetic group in the Blue Mountains represented what geneticists call 'adaptive radiations' from the dominant genetic groups. A bigger sampling effort encompassing larger areas would be beneficial to understand genetic drift and dispersal patterns of wild dingoes.

The population genetic structure of dingoes in the Blue Mountains showed some evidence of genes between groups being mixed although most had genes that were specific to their pack. These data are consistent with my suggestion earlier, giving weight to the idea that it is possible to identify the genetic relationships between individual dingoes and spatial relationships of dingo packs in the wild. Population structure of captive-bred 'pure' dingoes and wild dingoes has not been published previously. Investigation of relatedness between captive populations would be an effective test of dingo purity estimates because levels of inbreeding or outbreeding could be determined. Perhaps research on population structure of captive-bred dingoes will provide a good indication of relatedness among animals in captivity, or relationships between morphometric measurements and colour.

Fitting phenotype, purity and relatedness together

All of the information presented above is quite complex and interrelated in a way that can be difficult to grapple. Essentially, what it suggests is that if someone wanted to create a pure dingo or a hybrid, it could easily be done through selective breeding. Defining pure dingoes based on physical characteristics has become an unreliable method to infer purity or to select samples to test for genetic purity. Variation in colouration also was originally linked to hybridisation rather than to geographic variation. Table 3.2 shows colour variation across Australia where dingoes with tan colouration decreased from the north (Kakadu) to the south of Australia (Victoria), while black and tan colouration increased.

Table 3.3 alternatively compares skull measurements for dingoes, hybrids and dogs with populations from Queensland, Eastern Highlands of Victoria and the Blue Mountains. It is most interesting that each site has measurements that are more dog than dingo, more dingo than dog, and hybrid scores.

Table 3.2 Percentage coat colour for comparison of dingo populations around Australia. (Adapted largely from Corbett 2001; south-east Queensland data by Elledge et al. 2006)

Location	Tan	Black & tan	Black	White	Sable	Brindle	Patchy	n
Kakadu	99	0	0	0	<1	0	<1	500
Central Australia	88	5	0	4	2	<1	<1	1320
Western Australia	72	15	0	7	0	3	4	256
South-east Qld	18	10	5	–	17	1	5	56
Blue Mountains	23	38	0	0	31	0	6	47
Vic. Highlands	43	26	3	<1	14	7	6	734

Table 3.3 Comparisons of average canonical measurements of dingoes, hybrids and dogs from around Australia. Adapted from Corbett 2001.

	Corbett (2001)			AUG (n = 41)	WQ (n = 15)	SEQ (n = 5)	FI (n = 21)	EHV (n = 338)	BM (n = 10)
	Dingo	Hybrid	Dog						
X_1	25.1	22.1	20.8	26.4	25.7	25.4	26.8	24.4	25.6
X_2	60.3	60.1	62.8	62.0	60.7	66.7	63.0	63.1	59.0
X_3	7.5	6.8	6.8	7.8	7.6	7.8	7.9	7.2	6.8
X_4	9.5	9.4	9.8	9.7	10.0	10.7	10.4	10.1	9.4
X_5	33.5	30.3	28.4	32.8	32.7	33.2	32.8	34.6	36.0
X_6	11.6	10.7	10.2	12.6	12.5	12.4	12.1	12.4	12.1
X_7	55.9	55.2	58.2	57.0	56.6	57.3	56.9	58.8	59.3
X_8	54.6	49.8	50.5	55.6	54.9	56.9	55.6	53.0	54.3
SL	–	–	–	196.6	193.4	183.6	193.4	–	196.7

AUG = Augathella; WQ = Western Queensland; SEQ = South-east Queensland; FI = Fraser Island (Woodall et al. 1996); EHV = Eastern Highlands of Victoria (Jones 1990); BM = Blue Mountains (Purcell 2010); X_1 to X_8 represent skull measurements for 'purity' assessment (Newsome et al. 1980); SL = Skull length.

Two individuals that we recaptured in the Blue Mountains were tested for genetic purity and relatedness twice using a pseudonym to test the accuracy of the tests. Both individuals clustered with the same genetic group using the relatedness test and showed consistency. One individual, however, returned results that suggested a higher proportion of 'pure' genes were prevalent during 2006 (75% pure) when compared with 2005 (65% pure). Only 13 tests were run in 2006 and 19 tests were run in 2005, so this could be the reason for the observed difference. In contrast, however, the second individual returned a 'purity' score of 6 (less than 50% dingo) during 2006 and 2007, although 9 tests were run during 2006 and 20 tests were run during 2007. These examples provided an indication for accuracy of the purity test.

Other novel observations were made when comparing purity with coat colour and morphometrics. Dingoes that matched the coat colour criteria scored as less 'pure' than dingoes that did not match the coat colour criteria. Sable dingoes had purity scores of between 2 and 6 (see Table 3.4) and presented a higher proportion of 'purity' (5.3% of sable animals were 'pure'; $n = 1$) than tan, and black and tan animals. Only 25% of tan dingoes scored 4 while 16.7% of black and tan dingoes scored 4. Dingoes that were within dingo colour and morphometric measurements were classed as less than 65% or less than 50% dingo. Table 3.4 relates genetic purity with coat colours and

Table 3.4 Relationship of genetic purity scores with coat colour from dingoes in the Blue Mountains.

'Purity' score	1	2	3	4	5	6	7	Total
Tan *n*	0	0	0	2	3	3	0	8
Tan %	0	0	0	25	37.5	37.5	0	–
B+T *n*	0	0	0	3	8	7	0	18
B+T %	0	0	0	16.7	44.4	38.9	0	–
Sable *n*	0	1	0	4	8	6	0	19
Sable %	0	5.3	0	21.1	42.1	31.6	0	–
Patchy *n*	0	0	0	0	1	2	0	3
Patchy %	0	0	0	0	33.3	66.7	0	–

B+T = black and tan colouration

Table 3.5 provides a cumulative summary of dingoes which fall within or outside of colour, morphometric measurements and genetic purity criteria.

Other scientists from Queensland also compared genetic purity estimates with coat colour and canonical measurements, and showed that fewer and fewer individuals could be assigned a pure status using the entire 'purity' criterion. They tested for consistency between estimates, and showed 69.6%

Table 3.5 Relationship of dingoes from the Blue Mountains that do or do not match coat colour, morphometric and genetic purity criteria.

Dingoes which:	Site 1 (*n*=11)		Site 2 (*n*=37)		Total (*n*=48)	
	n	%	*n*	%	*n*	%
Match coat colour (tan, B+T or white)	8	72.7	18	48.6	26	54.2
Do not match coat colour (sable or patchy)	3	27.3	19	51.4	22	45.8
Match morphometrics	4	36.4	3	8.1	7	14.6
Do not match morphometrics	7	63.6	34	91.9	41	85.4
Match morphometrics and coat colour	3	27.3	2	5.4	5	10.4
Do not match morphometrics and coat colour	8	72.7	35	94.5	43	89.6
Match morphometrics, coat colour and are 'pure'	0	0	0	0	0	0
Do not match morphometrics, coat colour and are 'pure'	0	0	1	2.7	1	2.1

B+T = black and tan

of animals were assigned the same status using all three techniques, 17.9% were assigned the same status using genetic tests and skull allometry, 7.1% by genetics and appearance and 3.6% by skull allometry and appearance. Regardless of these tests, it is still speculative whether purity can be assigned to dingoes using any technique.

What can be done? Do we accept the current estimates for purity and hybridisation and cull anything that is outside set criteria? Or do we accept that current estimates are only estimates, and dingoes have been misidentified? Since we had the first results to determine genetic relatedness between dingo packs in the Blue Mountains and showed that there was a high level of relatedness between most individuals in packs, we looked at the colour of individuals within each genetically related group. Peter Thomson had already shown that certain colour varieties were prevalent in some packs more than others in WA, but he did not have the genetic data to support his supposition. The figure on page 72 shows that tan and sable colouration was dominant in the blue genetic group, sable and black and tan was dominant in the green (and red) genetic group(s), and patchy, sable and tan colouration was dominant in the orange genetic group. The orange genetic group was also significantly outbred and the other groups were neither inbred nor outbred.

These data show that nationwide colour variations could either be adaptations to or the result of local environmental conditions. The patchy colouration of dingoes in the orange genetic group is not a rare event either. In studies of coat colour variation in northern and central Australia, 4.2% of 388 dingoes were ginger with white spots and 0.6% had patchy ginger and white colouration. Evan Jones also reported 2.8% were patchy ginger or ginger and white in the Eastern Highlands in Victoria. In the current study, 6% of 47 captures were patchy and all of the patchy coloured captures were from the same genetically related group. In all major studies that reported colouration of dingoes in Australia, patchy was not an unusual occurrence and therefore should be reconsidered in descriptions associated with 'purity'. In addition to this, it can be hypothesised from population structure data that limited breeding opportunities for the current population of Blue Mountains dingoes, due to eradication programs, may have resulted in coat colour variations. These data are also consistent with variation in coat colour of silver foxes that were bred within an intense selection pressure environment. If it only took six to eight generations for standard coat colour and skull allometry of silver foxes to vary, it may only take eight generations for colour and allometry of isolated dingo groups to vary. Results from the fox data showed phenotypic variations within 'pure' genetic stock. In respect to dingo purity, variations in phenotype may also arise under artificial and natural selection conditions. In fact, they have!

On 4 June 2009, the ABC science show *Catalyst* reported results from our study in the Blue Mountains on national television. Within eight days I received an email from a dingo breeder that stated the following:

> 'Here at our sanctuary, where we have … bred only from DNA tested "pure" dingoes, we have had quite a few pups born with heavy white spotting – up to 3/4 white collar, long white socks, white tail and even some large splashes over the midline. I admit that we select against these … and … cull soon after birth.'

How many dingo breeders select against colour varieties that appear to be hybrids? More importantly, if other colour varieties were selected for breeding, what colours would result? Are dingoes really only tan with four white paws and a white tail tip or is this only how humans breed them?

Following on from the association between genetic relatedness and coat colour, it was my presumption that genetic relatedness and genetic purity also would cluster together consistently. Results showed, however, that the most genetically 'pure' dingo in the Blue Mountains was related to the most genetically 'dog' dingo. In fact there was no consistency between genetic relatedness and purity estimates.

So what is a dingo?

If dingoes can't be characterised by their coat colour, skull shape or genetic purity estimates, then we are once again left to ponder what a dingo is. Chapter 1 showed that the word *dingo* itself was not recorded in the dialect of Aborigines to mean *dingo, per se,* and Aborigines had no concern for dingo purity. Indigenous Australians, however, had great concern for the function of dingoes in Australia's landscape as the top-order predator, Australia's wild dog, and for the balance of nature. For contemporary Australians, the story of the dingo is almost a metaphor for environmental degradation during European history in Australia and the lost natural balance. We are happy to label the dingo as an icon if it helps our economy or as an enemy if it negatively impacts on the economy. But the concern for the function of this higher order predator was almost forgotten.

If we want to be superficial about this paradigm, we can cull all the dingoes, mongrels, wild dogs and their hybrids based on appearance alone, for the preservation of this Euro-Australian dogma. One genetically pure dingo in my Blue Mountains study was sable, his total length being outside traditional dingo criteria, and so too his head length. The only reason we attained samples from him was because he had been shot by a local landholder after he had bailed up a kangaroo in a dam with other dingoes, and according to the shooter he was 'a hybrid'. In contrast, his canonical score also suggested that he was pure. The current measures for dingo purity obviously remain imprecise, and further research is required before conservation actions for the preservation of wild dingoes or culling of hybrid dingoes are implemented. Observed differences in colouration and skull allometry may be due to environmental influences but generally appear to be genetic. Average measurements from Kakadu National Park (KNP) for instance, suggested KNP dingoes were taller and heavier whilst Blue Mountains dingoes were shorter, longer and lighter. These slight variations could be attributable to latitude, climate and topography.

If dingoes require a description then we need to include an array of information about their primitive, communal-living, hypercarnivorous nature. Effects of colonisation and subsequent isolation and persecution for over two centuries pose potentially significant limiting factors to our understanding of dingoes. Current molecular studies do not adequately address the question of hybridisation in the wild due to use of reference material that may be biased by sample design, similar to research on differences in skull allometry. Therefore, dingoes should be characterised per site or region based on minor genotypic and phenotypic variations observed. As a primitive canid, however, dingoes should have only one annual oestrous cycle. Based on these biological characteristics the definition of a dingo should be redefined, as follows:

> The dingo is a primitive, communal-living, hypercarnivorous member of the family Canidae from Australia, with one annual breeding cycle, and minor genotypic and phenotypic variations across their geographic range.

This definition will be used for the remainder of this book and based on information in the current chapter, the term *dingo* is all-encompassing of wild dogs and hybrids. Feral dogs are excluded, being any wild-living domestic dogs. Presumably, if we can maintain the primitive, communal-living, hypercarnivorous nature of dingoes, then we can maintain their functional role.

4

DINGO CHARACTERISTICS AND BIOLOGY

General descriptors for dingoes include erect ears, a bushy tail and ginger/tan, black and tan, black or white coloured coats. White points may be present on some of the paws and the tail tip, but not always. Dingo × domestic dog hybrids were originally described as having dingo-like appearance but coat colour may vary to brindle or patchy; however, there were reports that stated some dingoes with 'pure' skull allometry may also be ginger with white spots, patchy ginger and white, or brindle. Most researchers and members of the public believe that these colours, along with sable and patchy black and white, are indicative of hybridisation; but the original dingo hybridisation experiments showed that sable colours were no indication of dog ancestry and reported observations of black and white dingoes from settlers in Western Australia during the mid-1800s.

Including data on dingoes collected from the Victorian Highlands, Eastern Highlands of Victoria, central Australia, Kakadu National Park and the Blue Mountains, compiled average measurements of dingoes show that their head length is 22 cm, ear length is 10 cm, shoulder height is 57 cm, hindfoot length is 19 cm, tail length is 33 cm, total length is 126 cm, and

Table 4.1 Maximium and minimum recorded measurements and weight for male and female dingoes.

	Males		Females	
	Min.	Max.	Min.	Max.
Head length (mm)	210	256	202.9	245
Ear length (mm)	89	112	89.35	104.1
Shoulder height (mm)	510	650	514	595
Hindfoot length (mm)	163	198	173	192
Tail length (mm)	160	570	346	530
Total length (mm)	1145	1540	1159	1465
Body weight (kg)	12.5	21.44	9.5	20.5

they weigh approximately 15 kg (Table 4.1). Despite popular belief that dingoes do not bark, CSIRO scientists have stated that dingoes *do* bark when alarmed, but otherwise communicate by howling.

Dingoes were on the Australian National Kennel Council website in 2007 as a breed classed in group four with hounds. The council stated that dingoes were a medium-sized and elegant breed that are also alert, rangy and instantly reflexive, with minimal excess flesh, and great agility and stamina. Coat colour was reported to range from red, ginger, gold to palest cream, and black and tan. Males stand 52–60 cm high and females less while displaying more femininity, with weight for males and females ranging between 13.5–19 kg. The Australian National Kennel Council also stated that male dingoes should have two normal, fully descended testicles and, despite information reported by CSIRO scientists, dingoes do not bark.

Whelping, rearing, training and breeding seasons

Dingoes generally have four biological seasons per year: breeding, whelping, rearing and training (see Figure 4.1). Breeding season usually commences around March and concludes during May, but this can vary between regions and individuals due to factors such as their age. Pro-oestrus (the period immediately before oestrus in most female mammals)/oestrus in captivity was recorded to last for 10–12 days but behavioural data from Western

Breeding (Autumn)				Whelping (Winter)				Rearing (Spring)				Training (Summer)		
Mar	Apr	May	•	Jun	Jul	Aug	•	Sep	Oct	Nov	•	Dec	Jan	Feb

Figure 4.1 The annual biological seasons of dingoes.

Australia suggested that it can last 30–60 days. Former CSIRO scientist Peter Catling once reported that male dingoes in central Australia were almost lacking sperm outside of the oestrous cycle except for when they were housed in a cool Canberra climate and they produced spermatozoa all year. Proximity to domestic dogs on heat in Canberra may have affected testosterone levels of dingoes in that study. Gestation periods observed in captivity in Australia lasted 61–69 days. Observations of dingoes at the London Zoo during the 1960s reported that pro-oestrus lasted two to three weeks, oestrus lasted three to four days and four recorded gestation periods were 59, 62, 62 and 73 days. Copulatory ties were observed in the northern hemisphere in October and as late as December, possibly as a result of crossing the equator. Litters in Australia are usually born in winter (June–August) confirming that dingoes are short day-length breeders.

There are contradictions in reports, however, when purity is being estimated for populations based on the time of year that whelping occurs. Litters from pure dingoes have been recorded in most months in central Australia, but if litters are born between November and April in south-eastern Australia, the female is presumed to be a hybrid. Dingo seasons observed in Western Australia included pre-breeding (14 Feb–22 Apr); breeding (23 Apr–25 June); nursing one (26 June–14 Aug); nursing two (15 Aug–6 Oct); post-nursing (7 Oct–4 Dec); and non-breeding (5 Dec–13 Feb). Most dingo researchers agree that the dingo breeding season is between March and May, births generally occur between June and August, and September–February is the non-breeding or the pup-rearing season.

Lactating female dingoes are thought to recycle water by coaxing pups to urinate and defecate through licking of their genitalia. Ingesting faeces from the pups by the mother is also thought to be a mechanism for developing immunity to infectious agents that pups may be in contact with, because the mother may develop antibodies and return these to the pups via the milk, much like epigenetic marks.

Social interactions and communication

Dingoes can hunt or navigate alone and within groups. Dominance hierarchies, similar to that observed in wolves with a dominant pair, subordinates, juveniles and omega (lowest ranking) animals are prevalent in captivity and in the wild. Communal nurture, where subordinate animals in wild groups do not breed but assist to rear the offspring of the dominant pair, is apparent but somewhat speculative. For instance, Laurie Corbett observed infanticide in captivity, when a dominant female moved her pups into the same den as the pups borne by the subordinate female. The young borne by the subordinate eventually died of starvation while both females suckled the young of the dominant female. In theory, the dominant female could require a subordinate female to be impregnated each breeding season so that the subordinate produces milk and the dominant female has assistance during lactation. This would act as population maintenance by regulation on one hand, because excess pups die, and increase survival of the dominant female and her apparently more genetically fit young, on the other.

During his research in Western Australia, Peter Thomson observed alterations in the function of social systems of dingo packs after an increase in population density due to an increase in pack size. These adjustments coincided with a reduced food supply after a period of little emigration and included increased emigration, disintegration of some packs and changes in territories, including some that contracted into favourable areas. Infanticide or population regulation by dominant females was alleged to occur. In other instances, multiple litters within packs were reported. This was hypothesised to allow rapid population increase following high mortality from events such as eradication programs. Wolf biologist David Mech also showed that the percentage of pups in a population between the period of natural regulation by the pack and artificial control by humans increased from 20% when wolf control started in 1951–52, to 35% in 1952–53 and then to 55% in 1953–54. Both David and Peters' observations are consistent with the behaviours discussed in Chapter 5. If resources are plentiful then the pack may increase in size until it has reached the upper limit of their resources, the time at which changes to social systems may occur.

Expression and postural communication

How dingoes determine that they have reached the upper limit of their resources is yet to be studied. It is common knowledge, however, that dingoes are communicative canids. Scientists have hypothesised that postural, vocal and olfactory communication can: a) maintain or increase distance between individuals; b) allow individuals to assess fighting or resource-holding potential of competitors; and c) provide a means to identify the location or movements of pack members and neighbouring groups. This is important when prey or water resources may be more abundant in the interstices of territories and resources are shared. It is also an essential trait when hunting larger game. During my field work, we disturbed a dingo from a hide near open grassland where large mobs of eastern grey kangaroos were feeding and frequently fed. This individual ran to more protective cover and began howling. Another four dingoes were then observed approximately 200 metres behind the original location of the first dingo and proceeded to flee at the sight of the vehicle. Photographic evidence from a motion-sensing camera in the area showed all five animals travelling together before the presumed hunt was interrupted.

There were many other occasions when dingoes responded to our presence vocally, yet two stand out. On the first occasion, a National Parks ranger and myself were camping on the Jenolan River because we had located a signal from a VHF telemetry collar that indicated the tagged dingo was deceased, and we were going to retrieve the collar and the skull. A dingo howled while we were setting up camp so I responded with a howl that the dingo responded to again. Then after dark, we were sitting around the camp fire and heard a dingo 'snuff' behind us. Laurie Corbett suggested that snuffing may help the dingo rapidly identify the intruder by increasing the intake of fresh air containing their scent, and that snuffing by wild dingoes had not been previously recorded. We quickly put a torch on the inquisitive ginger dingo standing 5–10 m behind us and it then slinked off into the darkness. Later I was woken in the early morning by a couple of dingoes howling near our camp site.

On the second occasion while in our field site with wildlife photographer Shannon Plummer, we happened upon a dingo pack at a fresh kangaroo kill and three of their pups. After taking some photos of these pups in a log, we were waiting on the fire trail just on dark when a dingo came trotting along the road. After it had seen us, it slinked behind us into the protective vegetation and started bark-howling frequently. Whilst only speculating, it seemed as though this form of howling was to warn other pack members of our location or to deter us from their core area. Responding with my own howl, the dingo then moved closer to us and continued to bark-howl and then the howls slowly drifted further away after I had ceased howling back.

The most visible signs of dingo communication are through facial expressions, positions that the tail is held and through full body expression. Facial expressions include: a) normal calm expression; b) anxiety, where the ears are held back but pointed upwards and the lips are drawn backwards; c) threat, where the ears are erect and point forwards and the mouth is snarling with teeth exposed; and d) suspicion, where the ears are held back and flat, the mouth is closed and drooping and the eyebrows appear crossed. Tail positions can include: being held upright, showing self-confidence; upright and slightly crooked showing suspicion of a certain threat; a normal flaccid tail or a flaccid tail with a slight upwards curve at the end for a not entirely certain threat; and even the classic tail between the legs showing submission and strong restraint. Tails can also be in between expressions to show moods such as depression or happiness.

Full body expression is usually best observed when subordinate animals behave submissively to dominant animals (Figure 4.2). When one of my pet dogs is in trouble, he generally rolls straight onto his back due to the tone in my voice and looks for approval to continue participating in any activity. Dingoes all react in similar ways when the dominant pack members assert their dominance over lower ranking pack members. British wolf behaviour expert Shaun Ellis has made unique observations on communication between wolves while living with them. One method of communication was as simple as noticing how the wolves leant against each other, and against him. New research on dingo behaviour and cognition is also currently being conducted. Rob Appleby, an ethologist studying dingoes on Fraser Island suggested to me that dingoes may sometimes even communicate by the twitch of an ear. He was watching two pups rumbling when they stopped and eyed each other off from a small distance. Then one pup twitched one ear slightly and the rumble was back on!

Having observed video footage from research on captive dingoes in South Australia puts dingoes in an entirely new category of clever. In one video, a dingo put its snout through a hole in its gate, unlocked the latch and walked free with the others. In another, the dingo moved a table in the pen closer to the fence where a treat was hanging out of reach. Then the dingo jumped atop the table, reached up and forward and pried the treat from the fence for a quick snack. Results from these studies will most likely reveal even more complexities about dingoes.

Figure 4.2 Submissive behaviour by a subordinate female when approached by the dominant female from a pack on Fraser Island. Image: Nick Alexander

Olfactory communication

Similarly to wolves and most other canids, dingoes rely on their sense of smell. It is not so much that their eyesight is poor, although it relies primarily on movement, but their noses are excellent. The olfactory system in a dog's nose is 14 times larger and up to 100 times more sensitive than a human's nose. Dingoes and other canids use their nose to identify individuals that they know and to identify other information such as where pack members have been and maybe even what they have eaten. For instance, my two pet dogs often won't sit still or stop jumping until they have smelt my breath. They know who I am obviously from my voice, physical appearance and my natural scent, but my breath appears to be the most important thing for them to register. This is most probably an instinctual behaviour, developed over millions of years. The scents on the breath of fellow pack members may indicate if that animal has food for the rest of the pack or if it found an interesting scent that the other pack members need to familiarise themselves with. David Mech discussed many occasions when wolves greeted one another by licking and smelling each others breath. During another observation, he said that he once watched a pack of wolves detect the scent of a large moose *Alces alces* upwind. The group then stopped, assembled closely, wagged their tails, touched noses together and then proceeded in single file upwind toward the moose.

It is relatively common to see pet dogs and other canids rubbing their neck, cheeks and chest or even completely rolling in a scent that they have found. This is most likely to advertise a new smell or perhaps even to mask their own smell if they are lone dingoes. On one occasion during fieldwork, I captured an image of *Mullunga*, a female in my study that we always saw alone, rubbing her neck in the attractant that I had made (Figure 4.3). Speculations about this form of behaviour are numerous but, if she was a loner, or an omega member of a pack, then that smell may have made her seem more attractive for other pack members or it may have confused other pack members into thinking that she was a higher ranking individual. Dog trappers go to great lengths to make a scent that is attractive enough for dingoes to lift their head off the scent on the ground so they don't smell the disturbed soil or the trap that they have set in their path. Usually when a dingo detects a foreign scent, it will refresh that scent post with its own scent to re-mark its territory.

Figure 4.3 *Mullunga* rubbing her neck in the attractant.

Diet and nutrition

There does not appear to be any data on daily energy requirements for dingoes. They consume approximately 7% (≈1 kg) of their body weight in food per day and in desert regions about 70–100 ml/kg per day in water, though it is important to remember that fluids may also be absorbed from prey. On one field trip I filmed dingoes kill a kangaroo, eat the genitalia and surrounding soft tissues and then they appeared to be drinking the blood, presumably from the ephemeral artery inside the thigh. There was an abundance of water in my study area, so apart from the flavour, drinking the blood may not have been necessary. Dingoes in desert regions may, however, supplement some of their water requirements by drinking blood from prey.

Major prey for dingoes consists of mammals, birds, reptiles and vegetation but varies depending on prey abundance and prey availability. Dingoes generally exhibit a combination of opportunistic and selective predation that may be relative to prevalent social systems. It is safe to assume that changes in pack structure and hunting strategies may alter the composition of prey dingoes can catch, or there may be variation between individuals based on their status in a pack. For instance, dominant animals may be first to feed, consuming gut contents of the prey while low-ranking animals may be forced to scavenge any available carcasses or prey. Dominant dingoes may therefore show higher vegetation content in stomach or faecal samples, while low ranking animals may show a higher incidence of fur or bone fragments from scavenging. These ideas may not yet be testable in field studies but are within the realms of dominance hierarchies. Similarly, neighbouring packs may display dietary differences based on prey availability within their territory.

Dietary studies of dingoes from eastern, central and Western Australia have displayed dietary differences across Australia based on prey availability within their regions. From analyses of stomach contents from dingoes in Western Australia, 40% of prey consumed was the euro *Macropus robustus*, 26.9% was the red kangaroo and 5.5% was the introduced European rabbit *Oryctolagus cuniculus*. Dingoes consumed dingoes as much as they consumed introduced sheep *Ovis aries* (4.1%), which were speculated as 'probably non-feral', which suggests that there were feral sheep in that area. Ground matter was observed in 54.5% of stomachs, birds were observed in 15.3% and remaining contents ranged from large, medium and small native and introduced mammals (11.2%), reptiles (5.6%), insects (5.5%) and miscellaneous matter (1.4%). It was concluded from that study that dingoes were an opportunistic predator despite the fact that little introduced domestic livestock were represented in the stomachs and they were as common in the site as the euro and the red kangaroo. Similar observations were made in the 1970s and 1980s: 96% of food items were mammalian, significantly dominated by rabbit and red kangaroo with minor insignificant contribution of cattle which was available as carrion. Peter Thomson also reported red kangaroo as the staple prey in Western Australia, even when cattle carrion and sheep were as abundant in his study area.

Analyses of stomach contents from dingoes in central Australia suggested that dingoes switched between prey items, depending on prey availability. On the whole, rabbit was most commonly eaten (56%), followed by small mammals (27%), cattle (16.8%) as prey (2.1%) and carrion (6.3%), red kangaroo (15%), lizards (12%), birds (4%) and insects (2%). When the dingoes alternated between prey species they were observed to predate over-abundant rodent populations for one year at the cessation of a severe drought. European rabbits then increased as items of diet for three years during wetter periods when they accounted for 69% of dingo diet in the first flush and up to 90% in the second flush. Dingoes were therefore suppressing the rabbit population. Following decline of the rabbit population in the proceeding drought, red kangaroo increased as a prey item, as did cattle as carrion when they began to die off due to dry conditions. Other researchers have shown more recently that rodents (42%), macropods (39%) and rabbits (29%) were the most common prey items and that the occurrence of rabbits as prey directly correlated with rabbit abundance in their study. In a site further north, reptiles were shown to be a staple prey item for dingoes.

Many more dietary studies have been conducted in eastern Australia so they are represented in Table 4.2. Research areas for these studies have ranged from south-east and north-east Victoria and south-east, east and north-east NSW. In general, all studies were linked through the Great Dividing Range.

Additional research in Gippsland during the 1970s showed a high incidence of possum (brushtail/mountain and ringtail), swamp wallaby, bush rat and rabbit but did not separate scats of dingo or the introduced red fox. Other prey items reported within dietary studies are significant contributions to diet when combined but were relatively insignificant on their own. It is easy to see, however, that there can be large variations in dingo diet across Australia, and most do not include a high proportion of livestock.

Table 4.2 Dominant prey observed in dietary studies for dingoes in eastern Australia.

Site	Date	Major prey	%	Source
From mountains of Gippsland, NE Victoria and SE NSW and the SE NSW coast	1969–1975	Wallabies [a] Common wombats[#] Possums [b] European rabbits	50.8 12.9 11.6 7.8	Newsome et al. 1983b
Nadgee (SE NSW coast), and Kosciuszko (SE NSW mountains)	1971–1980	Wallabies [a] Possums [b] Water birds [c] European rabbits	41.1 17.4 36.9 10.2	Newsome et al. 1983a
Bega, coastal SE NSW	1981–1982	Wallabies [a] European rabbits Possums [b]	46.0 17.5 8.2	Lunney et al. 1990
GBMWHA, E NSW	2002	Wallabies [a] Possums [b] Antechinus sp. [d] Avifauna	37.1 21.3 11.1 9.5	Mitchell and Banks 2005
GBMWHA, E NSW	2005–2007	Swamp Wallaby Possums[b] Eastern Grey Kangaroo	39.8 12.2 6.29	Purcell 2010
NE NSW	1969–1974	Wallabies [a] Possums [b, e] Bush rat*	41.7 11.3 12.2	Robertshaw and Harden 1985

[a] Includes swamp wallaby *Wallabia bicolor* and red-necked wallaby *Macropus rufrogriseus*
[b] Includes common brushtail possum *Trichosurus vulpecula*, common ringtail possum *Pseudocheirus peregrinus* and indeterminate possum
[c] Includes little penguin *Eudyptyla minor*, muttonbird *Puffinis* sp., Swan *Cygnus atratus*, Eurasian coot *Fulica atra* and indeterminate large water birds
[d] Includes brown antechinus *Antechinus stuartii* and dusky antechinus *Antechinus swainsonii*
[e] Does not separate common brushtail possum or mountain possums *T. caninus*
[#] *Vombatus ursinus*
[*] *Rattus fuscipes*
GBMWHA = Greater Blue Mountains World Heritage Area

Activity and abundance

The most commonly used techniques to estimate activity and abundance of dingoes involve systematic sampling of areas where their signs are most easily detected. Sign can include direct sightings, hair, scent posts and scats, but footprints are used most commonly. Research objectives for studies on activity or abundance of dingoes in Australia have been to test the efficacy of techniques to monitor the population to determine the success of control programs.

Numerous researchers have tested the efficacy of various indices to investigate changes in dingo abundance or activity. Results from studies in south-west Queensland showed that different indices were sensitive to variation between individuals (age/pack status) and environmental constraints. In another study, researchers tested the efficacy of aerial baiting of a wild dog population in the north-eastern tablelands of NSW. Results indicated that levels of activity and abundance varied at each site, before and after baiting. A decline in activity and abundance was also observed at the site without treatment. Sites were a maximum of 12.7 km apart, within the scope of one dingo's home range, and shared a common boundary in Guy Fawkes River National Park that may have been the link between the two sites. These data indicated that 'targeted' aerial baiting campaigns may cause landscape scale impacts to dingo populations and the functioning of ecosystems.

Variations in dingo track densities were observed pre-baiting and post-baiting at baited and unbaited sites for cat, fox and dingo tracks during trials of aerial baiting control measures in the Gibson Desert of Western Australia. The distance between sites was once again within the home range of dingoes in that study area and the dingoes that lived in the baited site may have been the same dingoes that lived in, or shared, the unbaited site. Other researchers have shown that track-based surveys at one site can, however, be one to three times higher than at other sites. Therefore the activity of individuals as detected when using track-based surveys is a function of population density, and population reduction may increase activity of individuals, sometimes causing the results to indicate that a high density population of dingoes is still present. Differences between sites may be based on resource availability and interspecific competition, intraspecific competition, spatial organisation and geography.

Activity patterns of dingoes are generally highest at dawn and dusk, low around midday and lowest during the night. In north-eastern NSW the activity patterns of dingoes were not crepuscular (peaks in activity at dawn and at dusk), nor were they nocturnal, indicating that the dingoes were cathemeral, or had irregular patterns of activity. In Western Australia, dingo activity patterns were crepuscular with periods of rest in the heat of the day. Dingoes were most active in March, least active in June and activity from August to December fluctuated. Behaviours such as raised-leg urination and howling increased in frequency between February and May, and marked the pre-breeding season of the dingoes in that study site.

It has been suggested that if researchers wish to identify changes in activity, then data for indices of activity or abundance should be collected during discrete periods of the annual biological cycle of the species of interest. In the case of the dingo, discrete periods could be breeding, whelping, rearing or training seasons. Using this method, most data on activity and abundance have showed increased activity in the breeding (March–May) and rearing seasons (September–November). These data are most often used to inform control programs and as a consequence, dingoes are baited during breeding and rearing seasons when populations appear to be in higher densities.

Movement, home range and dispersal

The most comprehensive data on movement ecology of dingoes are from studies in north-east NSW and in Western Australia. Bob Harden tracked five adults and four juveniles from 1970 to 1974 in NSW. Using triangulation and signal strength techniques, 4058 fixes were obtained over 515 days. Tracking was intensive, initially obtaining fixes twice daily and then every 15 minutes for 100-hour intervals. Results suggested that dingoes moved with two different patterns. The first, identified as searching movement, was characterised by high activity in a small area. The second was characterised by purposeful movements over a substantial area and dubbed exploratory movement. Home ranges varied from 4.3 km² (from 1063 locations) for a juvenile female to 54.8 km² (from 52 locations) for an adult male.

Adult males were observed to occupy larger home ranges than adult females in the study by Peter Thomson in the Fortescue region of Western Australia. Mean 95% home range estimates were 84.8 km² for the three adult males, 56 km² for the eight adult females and 159.6 km² for the three lone dingoes. Pack territories revealed a large degree of spatial separation between different packs and considerable overlap of individuals within packs. Territories were stable from year to year and encounters between packs were rare. The number of locations ranged from 71 to 448 over a minimum period of eight months, spanning four of six biological seasons identified for that study. Riverine areas were preferred habitat and core areas of activity randomly shifted between seasons. Pup rearing, however, appeared to be the most influential factor in movement patterns. Extensive short-term or exploratory movements were also observed in WA.

Data from WA suggested that it was uncommon for dingoes to travel large distances across territories, though one individual travelled 19.2 km in 7.5 hrs and another 17.2 km in 6.1 hrs. Since techniques employed by Peter Thomson incorporated aerial telemetry, compared with the ground-based triangulation techniques used by Bob Harden, observations of movement patterns cannot be directly compared. Laurie Corbett, however, reported that dingo packs sharing water resources in arid environments howled to inform neighbouring packs of their approach. This behaviour is expected to minimise conflict between, and injuries to, pack members and may explain why inter-pack encounters were rarely observed in WA.

Incidence of dispersal was highest in WA when population densities were at their peak and food supply was low. As stated earlier, whole packs were observed to disperse in some instances and in other instances lone dingoes were seen dispersing, whose chance of survival was comparatively slim. Vacant areas appeared to be a requirement for any dispersing dingoes. These were usually more available in areas where humans had been culling dingo populations and the staple prey of dingoes in that area, red kangaroos, had abundant feed. This created a 'no-win' situation: if dingoes were controlled near farmland, predator-free modified landscapes with abundant resources for herbivores were made available. Emigration to farmland by dingoes would subsequently increase since water, food and space for territories were abundant. Requirements for dingo control would consequently increase. If dingoes were not controlled near farmland, livestock loss may occur, but according to other scientists, overall livestock loss has been shown to increase due to control. Hence, land managers may create an endless need for dingo control or management by controlling dingo populations.

Observational learning

Canids learn behaviours through observation. Researchers have demonstrated how domestic dogs learned behaviours by watching their mothers, conspecifics (an organism belonging to the same species as another, in this case being another domestic dog) or a human equivalent. Studies have shown that asocial (individual) and social learning abilities contributed to increased fitness of individuals. Observational learning thus becomes the nexus for cultural transmission of behavioural traits such as predation on livestock, predation on native species or scavenging from human settlements.

The above observations imply that the dingo at Uluru in 1980 that was involved in the controversial murder trial may have learnt through observation that exploring inside a tent would result in the opportunity of taking food. Dingoes are known to become habituated to finding food items around camp sites and this may have been reinforced by campers willingly offering food items to dingoes, much like on Fraser Island. Scientists that have studied cognition in domestic dogs have showed that development of problem-solving behaviours involved complex social learning interactions and individual experiences. In contrast, wolves are naturally skilled problem solvers, highly social predators with well-organised social hierarchy who depend on other pack members to acquire behavioural traits. Another researcher once compared trainability, learning and intelligence of a malamute, a malamute-wolf hybrid and a wolf, and concluded that wolves have a 'duplex' information processing system of cognitive processing and instinct because of their wild heritage. The malamute, in that study, failed to unlock a latch after six years of watching the researchers enter and leave the enclosure. The malamute-wolf hybrid, however, performed the task within two weeks and the older female wolf unlocked the latch after watching the hybrid do it once. In addition, the hybrid, having received no obedience training, performed the obedience routines being taught to the malamute over six days, instantaneously. Indeed it seems as if wolves will learn behaviours from conspecifics or neighbours much faster than they would from humans. Since dingoes also are a subspecies of the gray wolf, and a hypercarnivorous, communal-living canid, we can expect dingoes to learn behavioural traits through observation (Figure 4.4).

Figure 4.4 *Makileiko* eyes off the motion-sensing camera whilst her pups familiarise themselves with the attractant.

The dingo *in situ* but *ex situ* – management and legislation

Based on movement patterns, Robert Harden made two points clear:

1 limited movements and small home ranges indicated control only needed to occur near or adjacent to affected properties; and
2 delivery of poison baits by fixed wing aircraft would be ineffective to target dingo travel routes such as creeks, ridge tops and fire trails.

The use of buffer zones, where dingoes are controlled in an area as large as the home range of a dingo with sufficient resources, adjacent to livestock properties, was also identified as a method to reduce depredation of livestock. Dispersing dingoes would presumably settle, not migrate onto farmland and be effectively controlled by managers. To maintain efficiency, rivers and creeks needed to be targeted and it was recommended that campaigns coincided with biological seasons when dingoes were known to be more active.

These methods for control indicated livestock losses could be minimised with continuous effort at specific times of the year. Dingo management prior to implementation of buffer zone management techniques was aimed at eradication but did not account for recruitment of dispersing dingoes from other areas. Settlers' use of exclusion fencing, land clearing, poisoning, trapping and shooting were responsible for the dingo becoming extinct during the 1800s in the majority of south-eastern Australia. Shepherding slowly became a last resort for farmers as professional dog trappers were employed by the state to target troublesome dingoes. Bounty systems to promote dingo control were effective from 1836 until the 1990s when research revealed they were subject to fraud and ineffective. Poisoning has included use of strychnine but is now generally confined to close regulation of compound 1080 (sodium monofluoroacetate) injected into fresh meat and manufactured baits. Apparent improvements in the technique of aerial baiting validated this approach as an effective method of dingo control, with ground baiting also a frequently used method. Public scrutiny of lethal control, public interest in preservation of dingoes, and the impact of poisonous baits on native animals has improved due diligence over administration of poisonous baits in recent times. Use of compound 1080, however, is banned in the USA.

Legislation has led to a paradox for historical and contemporary dingo management. The major dilemma is that dingoes are protected under Acts enabled to protect native species, yet are declared pests under Acts enabled to protect livestock enterprises. Legislation also differs between states and territories of Australia. In the Australian Capital Territory (ACT), dingo control requires a permit while in NSW, the state that surrounds the ACT, land owners are obliged to control dingoes as pests. Pure dingoes in Victoria, the state directly south of NSW, were even listed as threatened in 2008. There is no indication that discrepancies in legislation between states and territories will be changed in the near future. The case of the dingo is far from being the only one in the world where a wild canid requires control and conservation. A number of topics ranging from cultural prejudices, economics, development, politics, and improvements in animal husbandry, animal welfare and maintenance of biodiversity also need to be considered when attempting to manage carnivore populations for conservation.

Legislation within particular states and territories permits dingoes to be kept as pets. Aside from this, dingoes are found in most zoological preserves and within kennels registered by dingo preservation societies. There is considerable emphasis on maintenance of dingo purity in privately held and zoological collections, though this concept remains difficult to define and validate. Registration of dingo breeders and pure dingoes on a national register has been recommended, with education of the public and the agricultural sector on conservation efforts involving dingoes also important. Since hybridisation in the wild is seen as a key threatening process for dingoes, isolation of pure dingoes in large areas is suggested to have merit. Placing them on large offshore islands or fencing areas are two extreme conservation techniques that may maintain pure dingo stock in wild conditions. Such extremes might not be necessary, however, if improvements in management for contemporary wild dingo populations are made on mainland Australia, and they may not be completely natural either. Effective management of wild dingo populations will require a re-evaluation of the dingo in Australia and the industries this animal is reported to threaten. Holistic, best-practice management objectives will need to be included and address the cyclic conflict between land managers and dingoes.

5

HYPERCARNIVORY, SOCIALITY AND TERRITORY INHERITANCE

Broadly speaking, the dingo is the Australian wolf. Researchers once studied a dingo pack in captivity and described how the pack formed and operated with a dominant male, a dominant female, and sub-adults while excess dingoes were outcast. Outcasts increased in numbers following each breeding season and included a male, but not always the same male, and sub-adult females. The sub-adult females that were not outcasts were apparently shown tolerance by other pack members probably because they were in oestrus or lactating. The pups of these lactating females had been killed by the dominant female on all occasions, so the lactating subordinates assisted the dominant female to raise her pups.

In stable dingo packs, disputes are settled by asserting dominance opposed to fighting, though asserting dominance may look like an act of aggression. Interpretation of social systems of dingo packs living in the wild in varying conditions, from forests in the south-east to the Simpson Desert were, however, always related to resource availability. In Western Australia, shifts in dingo social systems also coincided with high population density and a reduced food supply; however, the composition of packs was unknown. This chapter therefore largely comprises a review of canid societies to show:

1 potential reasons why some canids live in groups;
2 how group life may affect their patterns of movement; and
3 the relationship between sociality (the tendency to form communities and societies), sociability (the disposition or quality of being sociable), genetic relatedness, pup rearing, and territory maintenance and inheritance.

Social systems of hypercarnivorous canids

Based on the dispersion of available resources within a territory, group size and contrasting habitats, scientists have implied that most carnivores have a flexible social behaviour. Societies therefore vary depending on the effects of selective pressures on different individuals, whereby they become social beings based on the ecological and social circumstances in which they are born. Two overarching questions asked by scientists to understand the family Canidae are:

1 why do some canids live in groups, while others do not, and what shapes their societies?
2 how do body size and associated energy requirements relate to prey size and home range size?

In a study of energy requirements of terrestrial carnivores, researchers developed a series of mathematical models known as the net rate model analysis to answer these questions. This analysis was based on correlations between carnivore body mass and their most common prey. The net rate model analysis showed that canids heavier than 21.5 kg could not be sustained by small prey, and shifted to larger prey with equal or greater body mass at this point. Based on the amount of energy required to survive, time spent resting and time spent hunting were identified as the behaviours which ultimately shape canid societies. Since hypocarnivorous and mesocarnivorous canids only require small- to medium-sized prey to meet daily energy requirements, they can forage opportunistically alone. In contrast, hypercarnivorous canids are required to live communally to sustain adequate metabolic needs for hunting and resource defence. The mean live weight of dingoes is less than 21.5 kg; however, as previously stated in Chapter 2, dingoes can be classed as a hypercarnivorous canid because they live in groups, eat more than 70% vertebrate prey and they hunt animals that weigh more than their average weight.

Resources that require defence for optimal foraging include food, water and habitat within an optimal space. Discussion of optimal space requirements, or home range size, can take two directions. If resources are randomly distributed through a home range, an increase in group size demands an increase in range size to retain an adequate resource base. But if resources are not randomly dispersed, home range size will not be related to group size and is more likely to be related to resource abundance within the optimal space. Concern regarding why some home ranges become smaller as group size increases has been raised, though it is unclear whether those researchers were considering the home range to be the expected size for the calculated energy required for the group or the area traversed daily by the group.

The area traversed daily by the group will be different to the expected or required home range size because the strength of neighbouring groups and their area traversed daily will limit opportunities for range expansion. Wolf ecologist David Mech stated that prey resources for wolves, such as deer herds, were generally found in the interstices of territories. Alternatively, in arid Australia, neighbouring dingo groups apparently howled when approaching a water source to minimise conflicts between packs. Therefore, wolves share resources with neighbouring wolves and dingoes share resources with neighbouring dingoes in territory interstices. If resources within interstices are rich and, by division of territories, shared by neighbouring groups, then the areas traversed daily by each group must contain optimal resources for that group. The canids in question should therefore be able to increase the size of their pack within their home range until conflict between conspecifics becomes intolerable.

Sociobiological research on other hypercarnivorous wild canids provides some insight into the sociobiology of dingoes. Generally speaking, large canids such as wolves are speculated to have a skewed sex ratio towards males, resulting from female emigration and retention of male helpers. Due to the period of pup dependency, competition between females for male helpers to secure resources is likely to be intense and drive the society towards polyandry, where a female may mate with more than one male in a breeding season. Research has shown that the body mass of the mother influences the mass of the neonate, overall litter mass and gestation length. Essentially that means that when the mean live weight of a female canid is more than 13 kg, the pups are larger and more dependent on adults and the litter size increases. This implies that a pregnant female needs to maintain her fitness during pregnancy to give birth to fit pups and to do that successfully, she requires assistance from males. Increased pressure on available resources while females attempt to maintain their body mass causes conflict between females to secure male helpers and is implied to cause reproductive suppression. Alloparental behaviour, where offspring are raised and cared for by individuals other than their parents, occurs as a result, and shared responsibilities force the hypercarnivorous canids to move and hunt in packs (Figure 5.1). Hunting in packs increases the chance of success when hunting larger prey, defending territories and resources from competitors and maintains energy requirements for most members of the pack. Other researchers have, however, provided an opposing interpretation of canid sociobiology. Using genetic biotechnology techniques, facets of canid sociobiology were shown to be adaptive to changes within their territory and related to resource availability. These findings indicated that canid societies can increase in

Figure 5.1 The circle of life. A pregnant *Makileiko* (right) and her now fully grown pups (from Figure 4.4) appeared to be hunting when they walked past the camera between 7 am and 9 am over three consecutive mornings in April 2006.

size until they reach the upper limit of available resources to sustain their pack in their territory, or to sustain the entire population.

As a hypercarnivore, dingoes need to live communally to predate prey species that weigh more than themselves. The social systems that form as a result of communal living then tend to dictate the movement patterns of individuals and packs. Numerous studies have shown that social structures in carnivores form by tolerating or not tolerating conspecifics. Mechanisms for tolerance are related to prey availability, dispersal opportunities (space), strength in numbers, vulnerability to predation and/or reproductive success. Benefits of group living generally include cooperative breeding, increased hunting success and protection of kills, territory and young from interspecific competitors. Disadvantages of group living are intraspecific competition and reproductive suppression. This appears to be why the social systems of the subfamily Caninae affect their patterns of movement within and outside their home range.

Effects of social systems on movement patterns

Stemming from the benefits of group living, one can expect movement patterns of hypercarnivorous canids to revolve around foraging, maintenance of territory, breeding and rearing young. Alternatively, if intrapack competition is so great that the costs of remaining with a group outweigh the benefits to fitness of dispersal (or the cost of death!), establishing a new range or joining a neighbouring group, analyses of movement patterns may reveal changes in sociality. Shifts in social systems of dingoes have previously been observed for dingoes in Western Australia when resources had declined. Therefore the benefits and disadvantages to fitness may vary with circumstances per site. A resource-saturated site, potentially similar to the Blue Mountains where there were no limits to water or food, may facilitate establishment of larger social groups of dingoes and delayed dispersal. Alternatively, a resource-saturated site may produce more closely related groups in discrete territories over a broader landscape, similar to the genetic relatedness data presented in Chapter 3. When populations or groups exceed the maximum threshold for the optimal space, intraspecific conflict will dictate death, dispersal or changes in social structure.

Since dingoes are a canid species with a mean live weight that is greater than 13 kg, a sex ratio biased towards males can be expected because males use less resources (do not bear young) and provide more assistance in territory defence and hunting. Consequently, emigration of female dingoes is therefore hypothesised to occur more frequently at certain times of the year because they do bear young and place additional stress on the social system (such as require helpers) and resources. Dispersal and transience in dingo populations can therefore be expected to be sex-biased towards females. Researchers are yet to corroborate sex bias in dispersal of individuals from large canid populations. Dispersal may also be discussed as though it was a choice, between disperse and try to survive or stay and be killed, because it may be largely dependent on resource availability.

Investigating movement patterns

Radio telemetry has previously been used to monitor movements of canids and other wide-ranging carnivores. Studies generally tested hypotheses associated with home range size, home range use and patterns in activity and movement, including dispersal. Some research integrated analyses of behaviour to understand mechanisms for observed patterns in movement. Gregarious behaviours have been implicated as stimuli for altering feeding behaviour and patterns in activity and general movement patterns.

Findings from behavioural research on coyotes showed that the home range of coyotes consisted of core areas where animals spent extended time periods resting or hunting. A third type of behaviour was also observed periodically when coyotes ranged to areas that surrounded or were adjacent to core areas. It was hypothesised that this behaviour was used in the re-establishment of territory boundaries, investigation of neighbouring groups or as a second form of hunting behaviour. Scientists studying the behaviour of Ethiopian wolves *Canis simensis* observed similar movements and they termed the pattern 'border patrol'.

An underlying objective in many aspects of biological research has been to determine patterns in resource selection by animals. Generally speaking, if an animal or group of animals can secure a core area with optimal habitat, which includes access to reliable water and prey or food resources, then their movements will revolve around foraging and the maintenance of the optimal habitat. This is often shown in results of home range studies though attaining sufficient data to assess patterns of optimal habitat maintenance can be time consuming. Core areas would usually be maintained by carnivores by using olfactory signalling and other behavioural gestures to caution or inform sympatric competitors of their state. Depending on the reproductive cycle and the ambition of an individual or group to secure resources in an optimal habitat, territories need to be defined and may need to be defended. Numerous researchers have investigated how canids define their territories using scent from urine, faeces and anal gland secretions. Residents within a territory have to ensure neighbouring and lone competitors can identify boundaries and are made aware that if they choose to enter an area, instead of evading conflict, then they choose to contest conspecifics for resources.

On these grounds, the rationale for cyclical movement patterns within a territory can be understood better. If canids maintain scent posts at regular intervals, neighbouring individuals or groups are kept informed of changes within social systems. There are numerous functions of scent-marking behaviours. Observations and possible functions of scent marking include:

- territory marking;
- repeatedly marking novel objects;
- differences between gender and age;
- laying trails;
- signalling alarm and/or dominance;
- identifying individuals or groups;
- attracting sexual partners; and
- producing 'priming' pheromones to influence reproductive processes.

Results from some studies have provided strong evidence, however, that scent marking has a territorial function opposed to functioning as a method of self-orientation for resident animals. In addition other researchers studying coyotes have observed:

- a higher frequency of scent marks around the periphery of a territory than in the centre;
- a higher frequency of scent marks by dominant and resident individuals than by lone intruders; and
- an association between areas where there were high rates of scent marking and high rates of intrusion by neighbouring coyotes.

Changes in the frequency of scent post maintenance could indicate opportunities to enter a core area and challenge residents for territory or to join a pack to secure resources and increase gene flow. Seasonal activity patterns and changes in feeding behaviour may therefore be related to seasonal changes in territory maintenance, patterns of movement and patterns of activity.

The Territory Inheritance Hypothesis

Numerous researchers have proposed and tested hypotheses to understand formation of territories and group living better. Several of these hypotheses include the:

- **Prey Renewal Hypothesis** where rapid prey renewal reduces competition between conspecifics and slow prey renewal increases territoriality;
- **Antikleptogamy Hypothesis** where territoriality is a response for males to control mate access;
- **Integrative Hypothesis** where food resources are sufficient to sustain several individuals in a territory;
- **Constant Territory Size Hypothesis** (CTSH) where sociality is based on between-year fluctuations in food supply and the carnivores go through a food availability bottleneck; and
- **Resource Dispersion Hypothesis** (RDH) where sociality in carnivores will vary based on the spatio-temporal distribution of resources within a year.

The **Territory Inheritance Hypothesis** (TIH), however, is my preferred theory to understand the nature of dingoes. The TIH largely involves the theory of increased fitness and the maintenance of the fittest genes. For instance, a female satin bowerbird *Ptilonorhynchus violaceus* selects her mate based on the size and quality of the nest that he builds and, essentially, the most convincing argument put forward during courtship. Her progeny will expectedly inherit that level of biological fitness and both the males and the females' genes will survive.

The TIH therefore considers:

1   the increased fitness of the original territory holders when they ensure that the territory will be inherited by a carrier of their own genes; and
2   the increased fitness of a sub-adult that remains at home and survives in an environment otherwise saturated by conspecifics.

Through the observations of a captive dingo pack, we saw that the TIH was at play. One male was outcast annually, and females were either tolerated or outcast. If these individuals were living in wild conditions then the pack may have behaved more like the dingoes in Peter Thomson's study. Some ranges contracted to favourable areas while dingoes from packs that split up searched for vacant areas to inhabit. Under the fitness model in the

TIH, a strong and healthy individual stands a good chance in competition for vacant territories. A strong and healthy individual, however, may also be suitable as a helper in a pack or in forming a pack with other fit individuals. Interactions between individuals will obviously always be dynamic and dependent on whether population size is increasing or decreasing, possibly due to a disease outbreak, and the rate of survival either to adulthood or as adults. In a dingo population that is subjected to control, however, it is highly probable that there will be numerous vacant territories and limited opportunities to mate or form groups. If populations are constantly expanding or forming groups, then the animals will invariably react under the theory of territory inheritance.

The first reaction will be individuals forming territorial pairs, followed by breeding at an increasing rate and decreasing territory size. Then the dingoes will increase group size and decrease territory size, which will probably make space for territories of dispersers. Depending on resources then, if dingoes lived in optimal conditions, we would expect to see more closely related groups living in discrete territories over a broader landscape. The genetic relatedness data from our Blue Mountains study clearly showed exactly this. Closely related packs lived in close proximity to one another over a broad landscape. This is most likely the result from historical baiting campaigns that reduced opportunities for mating and increased available space for territories. If dingoes were living in less optimal conditions, such as the desert where water is limited, then the survival rate presumably decreases and opportunities for sub-adults to inherit their natal territory increases.

Developing an understanding of individual movement patterns among kin is extremely complex. Besides the usual factors that influence movements like the composition of the group, behaviour of group members and local environmental conditions, there are many other interesting and interrelated questions identified by American scientists. These include:

1   do philopatric individuals (those that return to a specific location in order to breed or feed after migrating) have lower mortality rates than dispersers?
2   why are some offspring permitted to stay while others are not?
3   are individuals encouraged to stay, or do they only stay if they aren't forced out or motivated by other reasons?
4   to what degree should parents influence the fates of reproductively unproven offspring?

5   to what extent do siblings influence one another?
6   how do offspring that stay assess their chances or the risks at breeding?
7   how do parents or siblings assess the availability of habitat and reproductive opportunities away from their natal area?
8   what benefits are there to remaining or dispersing?
9   is inbreeding among philopatric relatives a concern?

Answering these questions will help us to understand the function of dingo packs better, assist us in managing them as a top-order predator and in understanding their inherent role in the landscape. To fully test the TIH, however, it must be studied rigorously through long-term comparative field research from birth to death wherever possible.

6

HOW DO DINGOES SEE AUSTRALIAN LANDSCAPES?

This chapter shows how we can interpret data on dingo movements to understand territoriality, patterns of movement and use of space. Recent developments in technology such as motion-sensing cameras and collars fitted with Global Positioning System (GPS) devices have enabled data to be collected without devoting countless hours during the cool of winter, the heat of summer, early mornings, late nights and entire days in cars or aircraft to obtain enough locations and observations that demonstrate dingo movements. What makes this new technology even more useful is that the data are in real time and can be tailored to suit the research questions.

Motion-sensing cameras

Monitoring all forms of animals with improved camera technology has become the way of the future. The BBC had huge success using elephants holding a disguised 'trunk-cam' to capture an intimate view of tigers in the jungle, and motorised 'boulder cam' to capture footage of a pride of lions in their den. Capturing footage of dingoes in the Australian bush without such elaborate cameras and in only a small amount of time is difficult. Motion-sensing cameras have, however, enabled animals to be monitored during all hours of the day and night. They come in two varieties, active and passive. Active camera traps send an infrared beam between a transmitter and a receiver. When the beam is broken, the camera takes a photo. Infrared beams in passive camera trap systems, however, detect the combination of motion and body heat in a wedge-shaped area radiating outward in front of the monitor that is inbuilt with the camera housing. I find passive camera traps to be more effective because there is a larger infrared field for detecting animals. Using motion-sensing cameras on dingoes or other carnivores, however, requires a few tricks to maximise the quality of the photos.

Trick 1 – Think like a dingo

First, never forget that dingoes are landscape specialists. If you look at how humans have constructed fire trails and road networks in reserves, or the way that animal pads (the tracks that are frequently used by wildlife and domestic stock) have formed through a landscape then you can start to see animal-movement corridors. With animal pads, there are often a number of paths that start at the top of a gully and merge into one at the end – that could be a prime location for a camera. Dingoes tend to follow these routes, which often also follow ridges, rivers and gullies, because they are the easiest paths to walk. Another way to think like a dingo is to imagine how their territories may be set up, remembering that they need protective cover, den sites, food resources and water (see Figure 6.1). If you set your camera up so it points directly at a water hole then you will most likely get photos of many different species. Depending on the size of your memory card, megabytes could be wasted on countless kangaroo shots. Sometimes they can be funny and scenic, but they are easy. Good photos of dingoes are exciting because they are rare.

Trick 2 – Find scent posts and make your own lure

Scent posts are those sites that dingoes mark with urine, faeces and scratch marks to define their territories (see Chapter 5). A domestic dog can find these locations rapidly because they smell dingo scent and leave their own scent mark (you can watch domestic dogs find scent posts when they're walking along a street) – and the domestic dogs can double as a lure!

Typical colouration expected for a dingo × domestic dog hybrid.

Family portrait for the Douglas Scarp pack (orange genetic group, see page 72).

Representation of genetic relatedness between individual dingoes in the Blue Mountains. Each column represents one individual dingo, and genetic similarities or dissimilarities between individuals are identified based on colour codes. For instance, a completely red column indicates that the dingo is from the red genetic group, and a column with red and light green colour codes indicates that that individual had a mixture of genes from the red genetic group and the green genetic group.

Physical coat colouration of animals from the red group (a) was black and tan and sable; the green group (b) was black and tan and sable; the blue group (c) was tan and sable; the orange group (d) was tan, sable and patchy; and the purple group (e) was tan, black and tan and sable. Comparison of colouration with genetic relatedness indicated coat colour is an inherited trait per group or geographic location and not related to hybridisation.

Animal carcasses are sometimes common in the bush – here we used a dead swamp wallaby, the favourite food item for dingoes in south-east Australia, as the lure for this black and tan dingo.

We saw this dingo drinking from a dam moments after we checked this camera and saw this photo, and it was so sick from mange that I managed to follow it at a distance of 10 m for a short while with my video camera without being noticed.

Photos from motion-sensing cameras can be used to corroborate other data. In the top image, a dingo checks the scent after crossing the sand plot. In the bottom image, *Kurre* (ears) wears a VHF collar used to investigate movement patterns. Only one of the dingoes crossed the nearby sand plot, which was used to investigate dingo abundance and activity.

Were there actually more than two dingoes? These pictures display some problems when using infrared cameras for scientific experiments and attempting to identify animals that are the same colour and shape. In the top picture, the front right foot is hidden and the back left foot is exposed, and in the bottom picture the front right foot is exposed and the back left foot is hidden. I presume that the top dingo was a female and the bottom dingo was a male, but there is still uncertainty.

The top image is a good example of colour variation within a pack, and of a camera that was pointing a little too low. The bottom image was taken from the same location, and is a photo of the ginger dog (*Jarrjarr,* in the centre of the top photo) moments after he stepped in the trap, and *Willinga,* a juvenile black and tan male, before he walked into the next trap five metres after *Jarrjarr*'s trap.

What is that smell? Some dingoes appreciated the lure so much that they spent 10 minutes smelling and rolling in it.

Helter skelter! Some pictures occur only by chance. This camera was constantly taking photographs due to the windy conditions when this dingo bolted past, running away from me driving along the road behind it. All that we saw, however, was this picture and fresh dingo tracks on a sand plot.

The top image shows that some dingoes are very cautious when investigating an unfamiliar scent. The bottom image shows that dingoes are swift and can react to a surprise, like a flash or a trap, in a matter of milliseconds.

Dingoes that smell the scent from invaders will investigate the smell and re-mark their scent post. If you don't have access to a domestic dog, then you can easily make your own lure by collecting domestic dog faeces and mixing sardines, off milk, cheese and yoghurt, old meat, fat, cooking oil, blood and eggs together or any combination of the above, and leaving it in a plastic bottle to ferment. We once trapped a dingo using the oil from a can of sweet chilli tuna! Once you have found a scent post, usually a defining object like a small bush, a rock, a log or an embankment, sprinkle some lure on the spot, set your camera up on a tree about a metre away and leave the vicinity.

Trick 3 – Use a road intersection (or where an animal pad intersects with a road)

Road intersections are hotspots for motion-sensing cameras. They are often frequented by dingoes because they dissect dingo territories and have multiple packs that walk past, including resident individuals and extraterritorial foragers. It is also important to face the camera along a road, opposed to perpendicular to the road, so you can see whether there is more than one dingo. Once I watched a pack of dingoes approach a dam and the camera missed dingo one (who had no concern for the lure), photographed dingoes two and three because dingo one tripped the beam moments before they were in view, and nearly missed dingo four. If the camera was perpendicular to the road, it would have missed dingo four, and probably only would have photographed the back half of dingoes two and three.

Trick 4 – Frame your shot

My favourite camera post was a tree growing right next to a road gutter. It was at a road intersection, in the interstice of two to three dingo territories, approximately 100 m from a waterhole. There was a lightly vegetated hill in the foreground and a cliff line in the background that often displayed spectacular colours at sunset. When framing your shot you are only limited by your imagination; however, be cautious that the camera is not pointing directly at the rising sun because on most occasions the light is harsher and it is difficult to identify the animals. Sunsets aren't so bad because they can provide a nice, soft light and silhouette the animals. Make sure that the camera is at an appropriate height to detect dingoes; about one metre off the ground, and perpendicular to the dingo's bodyline. If it isn't perpendicular you will photograph too much of the ground or too much of the sky and trees. After positioning a camera, I used to stand in the foreground and check when the infrared beam detected the motion from my arm that I waved up and down slowly. This gave an indication of the height of the beam, the size of the area that it was scanning and the direction that the camera was facing.

Other factors that you may wish to consider when purchasing cameras are whether the camera takes infrared (usually with a green or red hue) or colour photos, and if the latter, requires a flash at night. I have countless photos of animals looking directly at the camera, so they can obviously hear it turn on, especially feral cats. A flash at night time also startles animals and on occasions, the light from the flash refracted off the perspex cover and back into the picture. I personally prefer using a flash for colour photographs to assist identifying individuals, but I have also seen some well-composed infrared images. There is the option of purchasing a camera that can record video footage; however you may have to sacrifice available photo opportunities at night time. Motion-sensing cameras can also be effective at revealing what animal took the bait, how many animals were available to be trapped, population health, sociology, behaviour and patterns of

Figure 6.1 My first photo of a dingo from a motion-sensing camera placed at an intersection of roads, pack territories and near a river.

movement. In fact, motion-sensing cameras can tell small parts of a story over time. We watched a dingo pack grow, for instance, because they were photographed consistently from August 2005 through to April 2007. During September 2005 they travelled past the camera with pups at 17:30 every afternoon for at least one week. During April 2006 when they were a fully grown pack hunting kangaroos, they travelled past between 7 am and 9 am consecutively for three mornings, and posed for Figure 5.1 one morning.

Whatever you do, don't forget to check the batteries and update the date and time on the camera each time you use it. Digital images usually record the date and time, however, transferring files between computers and automatic time and date updates can change the properties of the digital image. Remember to be cautious when saving files and to be consistent with data storage.

There are many other trends that can be viewed through the eyes of hidden cameras. For instance, unknown individual dingoes were photographed most often during the breeding season, pups were photographed most often during the whelping season and packs were photographed most often during the rearing and training season. Dingoes also were photographed most often in mornings and afternoons, a crepuscular pattern of activity that is common among dingo populations in south-east Australia.

Home range and core areas

In 1943, naturalist William Henry Burt wrote the first and most comprehensive review of territoriality and home range concepts. William stated that behaviours associated with territoriality are highest among the human species, because humans consider it an inherent right to 'own' property and space. The entire globe is separated into political territories and includes some that are highly prized and defended due to the available resources. A territory is 'any defended area', the protected part of the home range, and home range is the area traversed by an animal in its daily movements over a defined period of time. A core area of activity is a selected area which contains home site, refuges and dependable food sources. These definitions, however, are ambiguous. The term *home range* implies that an animal is resident for a long period and may instead be defined as the area the animal normally travels during routine activities. Territory, however, could be defined as the area occupied exclusively by an animal and maintained by overt defence or advertisement, like scent posts.

Terminology aside, the scientific community simply needs to observe, compare and report on consistent and repeatable behaviours from their research subjects to assist interpretation of data. The words *home range*, *territory*, and *core area* may be interchangeable because maintaining scent posts on the border of a home range is a form of defence, and because core areas generally don't overlap, then a core area should technically be called the territory. What size core area, however, represents a territory? I used 50% core areas in my study but core areas could be as large as 60–70% of their home range size before the boundaries of core areas meet. The most important part is that the methods of analysis are either similar or the same so studies across Australia or of canids around the world are comparable. In the Blue Mountains, I defined a core area as the area where dingoes spent 50% of their time, inside the highest probability of occurrence, and their home range was the area that they spent 90–95% of their time. Movements outside of the home range were extraterritorial forays, and I used the term 'dispersed range' to categorise them as an irregular event in dingo movements. A dispersed range is different to a dispersal event because they are short and the dingo is philopatric and returns to its home range. Dispersing dingoes settle outside of their home range and either establish or attempt to establish a new territory.

Playing dot to dot

When data on the movement patterns of dingoes are complete, scientists get the pleasure of joining every 'dot' location with the next. That sounds much worse than it is because computer programs these days do most of that work for them. Understanding what the dingoes are doing from one dot to the next is always open to interpretation so standard methods of analysis are used to somehow level out the playing field.

The Minimum Convex Polygon (MCP) analysis, for instance, has been used most frequently for dingoes. This method is as simple as drawing lines between the outermost points of the home range and calculating the area inside the polygon. Figure 6.2 shows the MCPs for dingoes studied in the north-east NSW tablelands in the 1970s and 1980s.

From these representations of home range it is easy to see that the MCP method provides a very general picture of where an individual

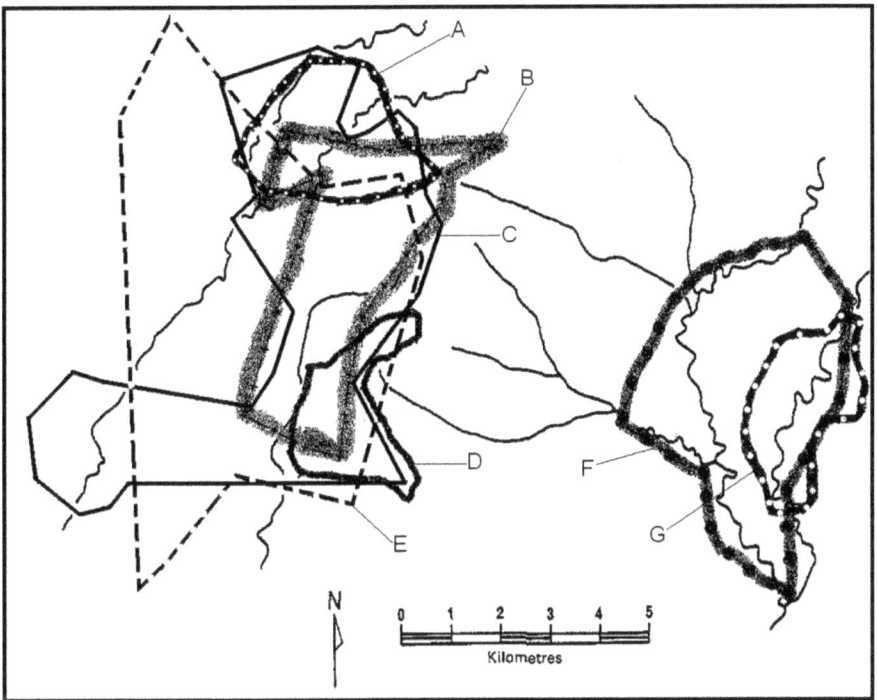

Figure 6.2 Representations of home range using Minimum Convex Polygons for seven individual dingoes in the north-east tablelands of NSW. Adapted from Harden 1985

lives. Each MCP can be made for different-sized areas, so the 95% MCP home range joins the dots around the inner 95% of locations and a 50% MCP core area joins the dots around 50% of the innermost locations. When the 95% MCP for individuals are compared with each other, then we can start to see who is a member of a pack and who is a neighbour. The 95% MCP analyses usually show areas of overlap between individual dingoes and neighbouring packs. These areas of overlap are generally referred to as the interstices between packs, and commonly contain shared prey and water resources. The 50% MCP core areas of individuals that live in the same pack, however, generally wholly overlap, though there is usually no overlap between 50% MCP core areas between packs. That is most probably because the 50% core areas contain the most dependable resources such as food, water and protective cover. There are other techniques, such as kernel estimators, that are more robust and provide a similar albeit more accurate and refined result though none are as accurate as mapping the raw data.

Patterns of movement

Movement patterns of dingoes are generally investigated as seasonal variations in home range size and core area size though they can also include territory maintenance, extraterritorial forays and dispersal movements, and daily, weekly or monthly forays to certain areas within their home range. Figure 6.3, for instance, shows seasonal variations in 100% and 50% MCP data for four dingoes in the Blue Mountains. The 50% core areas usually contracted during the whelping season and gradually increased during the rearing, training and breeding seasons in the ex-pastoral lease habitat. *Mirrigang* lived in a different habitat type to the other three dingoes and had a larger home range than the other females. This is most probably based on resource availability. It is important to note that the 100% MCP was maintained during the whelping season, although the 50% MCP core area contracted in size. This evidence suggests that males and females refreshed the scent posts on the outskirts of their home range and possibly monitored

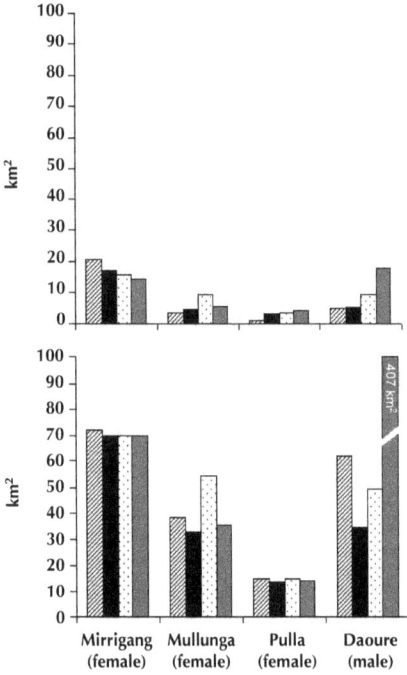

Figure 6.3 Comparison of 50% MCP home range estimates: (a) with 100% MCP home range estimates, (b) for seasonal differences (Whelping; Rearing; Training; Breeding).

the movements of individuals and packs walking past their territory during all biological seasons

Figure 6.4 shows the monthly movements of *Daoure* around his 50% MCP. In 2005, *Daoure* clearly reduced his movements during August, when pups may have whelped, and during the rearing season (September–November). Then his wanderings expanded during the training season (December–February). In the breeding season (March–May), however, he travelled extensively, possibly more than 120 km in one week, east from his core area through the Nattai River Valley, and almost walked out of the protected area. Then he turned and walked back to his core area via rivers, gullies and ridgelines and walked west. In April 2006 he travelled directly towards a property housing captive dingoes, and demonstrated the strong effects that breeding season pheromones can have on a male dingo. Fortunately for us he returned to his core area in time for the GPS

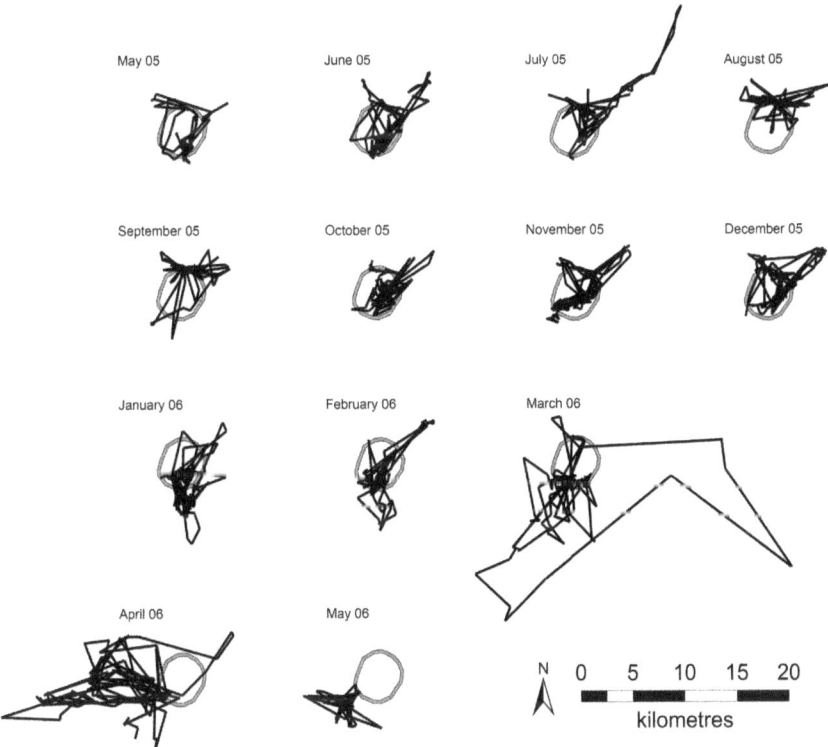

Figure 6.4 Monthly movements of *Daoure* showing site fidelity after extensive extraterritorial forays during the 2006 breeding season (March–May).

collar to release in an easy to retrieve location in May 2006. This behaviour begs the questions: If dingo pheromones were placed in the core of protected areas during breeding season, would dingoes migrate outside of protected areas in search of a mate or territory? Or would they travel to the centre of the park?

The dimensions of *Daoure's* walk, however, can't be seen in Figure 6.4. Figure 6.5 shows the cliff line that he traversed and descended when returning to his core area. The most plausible explanation for *Daoure's* navigation skills is that he followed scent posts and landscape features the entire way. We will never know whether he identified the Wollondilly River from the top of the cliff line, whether he was simply following scent posts or if he was travelling with an older companion that knew the direction to travel.

As you can see, so much data can be extrapolated from GPS collars that a whole book could be written using daily, weekly and monthly movements, and interactions between individuals and packs and landscape features. Figure 6.6a and Figure 6.6b are examples of *Mirrigang's* and *Pulla's* monthly

Figure 6.5 The cliff face *Daoure* negotiated during his extraterritorial foray and the peak (Bonnum Pic, once called Paddy's Peak – right off centre) that he descended when returning to his territory. Image: Robert Mulley

movements. Both females lived in contrasting habitat types (*Mirrigang* lived in forest and steep gullies and *Pulla* lived in ex-pastoral lease habitat); however, their forays were obviously reduced during June, when I suspect that they whelped. You may also notice that the GPS locations are reasonably evenly distributed north and south of their 50% MCP core area during each other month. It is important to investigate monthly movement patterns of dingoes in this manner because it shows that the females only breed once per year. Other researchers have hypothesised that hybrid dingoes and wild dogs may breed twice per year, exacerbating their impact on wildlife and livestock. If current techniques to identify dingo purity were less ambiguous and the Blue Mountains population were verified to be hybrid dingoes, then these data show that hybrids also only breed once per year and maintain a function in this landscape. Using the definition that dingoes should be characterised per site but defined nationally by their one annual breeding cycle and their function, these data and data from *Daoure's* movements demonstrate that the one annual breeding cycle is an integral, functional component of dingo behaviour and ecology.

Another technique used to investigate the function of dingoes with GPS collars is to assess how they maintain their territories and utilise landscape features. Figure 6.7 is an example of weekly movement patterns by *Jarrjarr* (sandy country) an adult male during the 2007 breeding season. The GPS collar that we fitted to him, and another that we fitted to the juvenile male *Willinga* (behind) caught beside him, were programmed to store one GPS location every 10 minutes. This was to see in finer detail how dingoes utilise landscape features, as a result of questions raised by *Daoure's* movements in March 2006.

In week 1, *Jarrjarr* travelled to his southern border and in week two he travelled to his northern border and then back to his southern border. In week 3 he first travelled north and then navigated part of his western border before returning to his core area. Week 4 he went to the southern and north-western borders, week 5 he wandered north and in week 6 he circumnavigated his home range. First he travelled north and then followed a ridgeline west and in a semicircle pattern, back to his southern border. Then he travelled back to his core area, and walked straight to the northern border. In week 7 and week 8 he wandered to and between borders. *Willinga* followed *Jarrjarr* on most of his border patrol movements and the pair may offer an example of an adult dingo training a juvenile dingo in territory maintenance. Another interesting note from *Jarrjarr's* circumnavigation of his territory was the rate of travel, and that it coincided with a full moon.

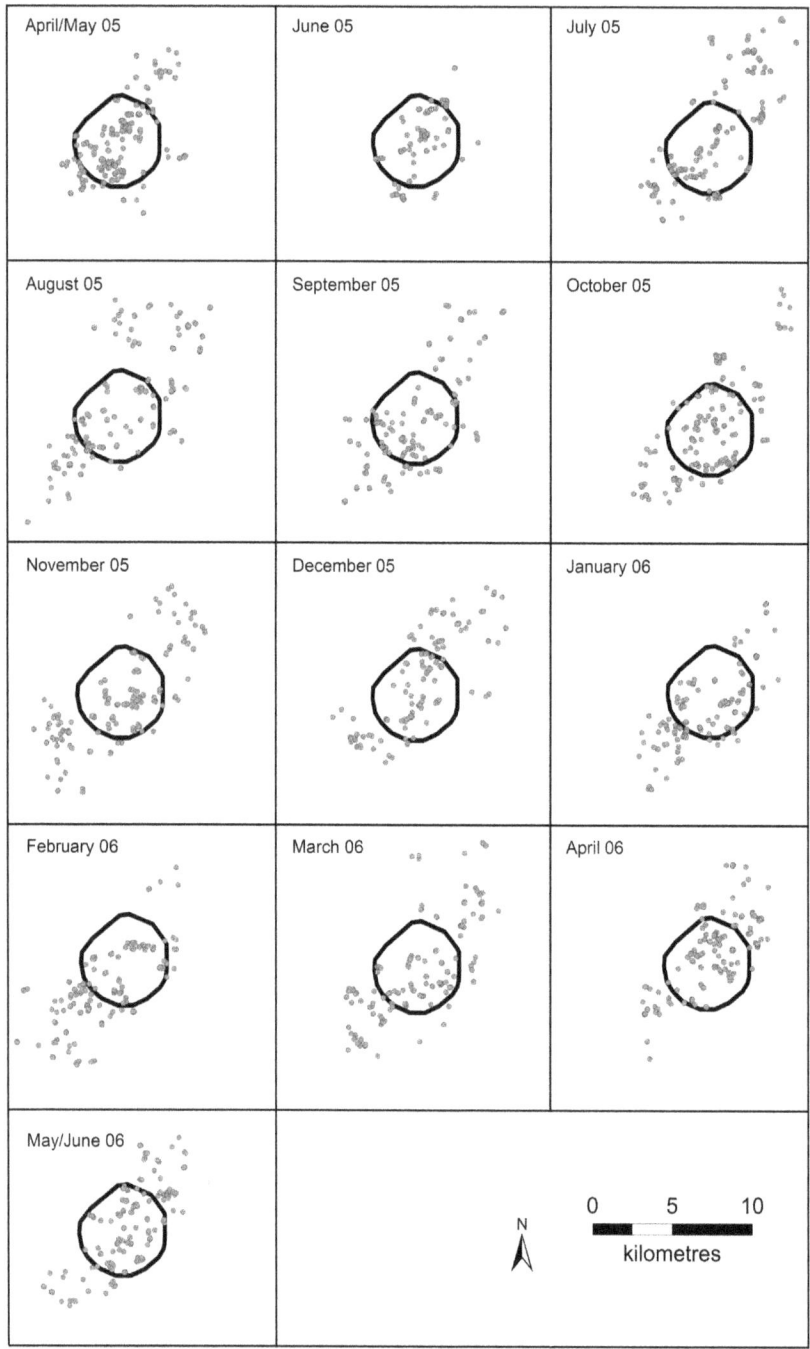

Figure 6.6a Monthly movements (each dot represents one location from the GPS collar) of *Mirrigang* around her 50% MCP core areas. Restricted movements inside the 50% MCP core area during June 05 are an indication of the month that she whelped.

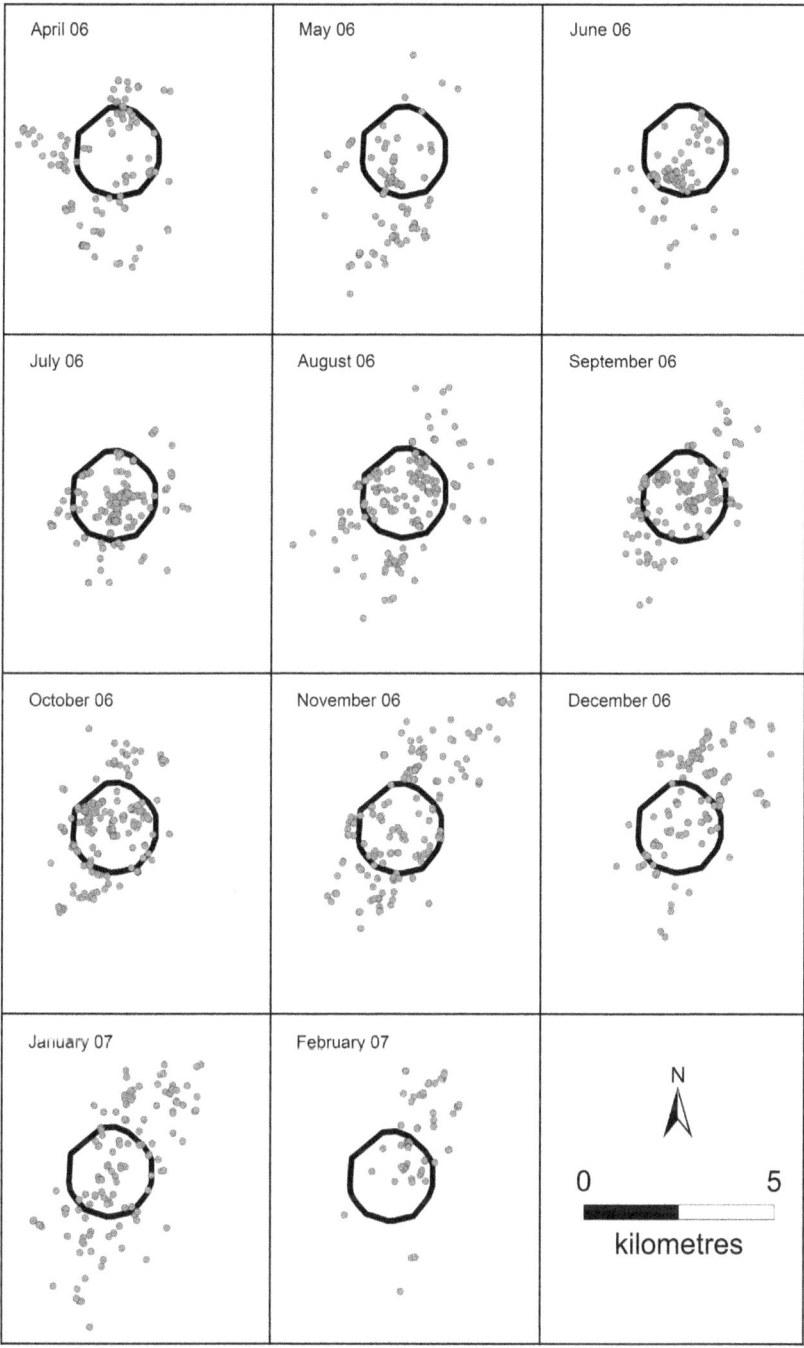

Figure 6.6b Monthly movements (each dot represents one location from the GPS collar) of *Pulla* around her 50% MCP core areas. Restricted movements inside the 50% MCP core area during June are an indication of the month that she whelped.

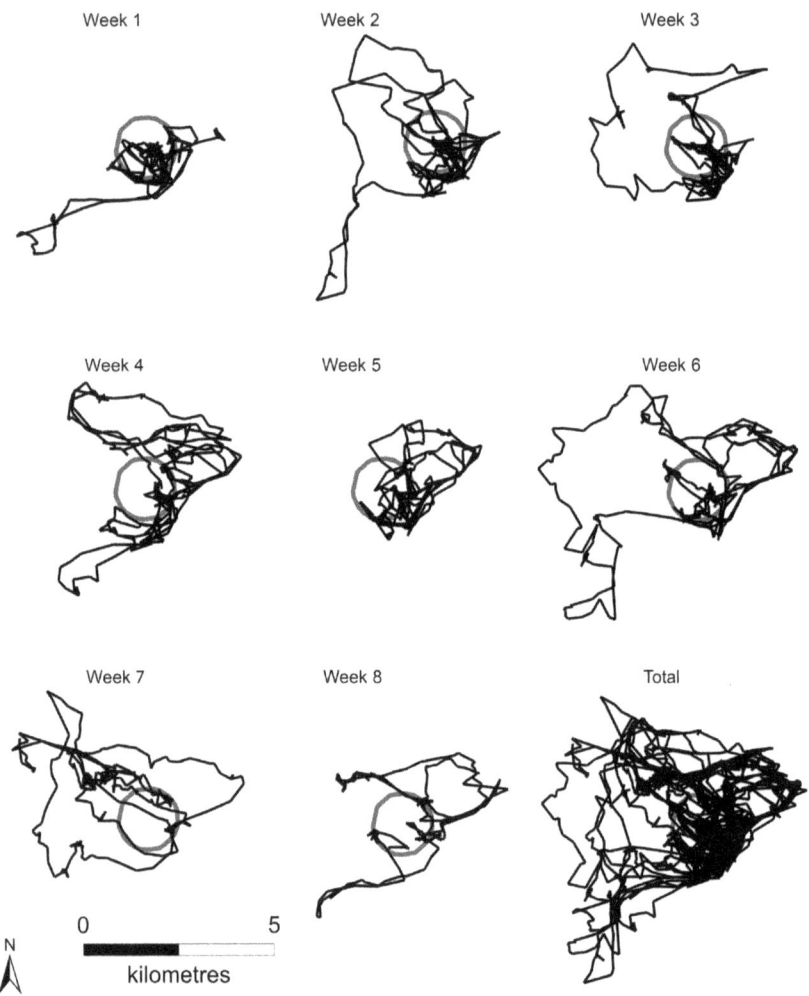

Figure 6.7 Weekly border patrol movement patterns (lines) by *Jarrjarr* (sandy country), around his 50% core area during the 2007 breeding season.

In Chapter 4, I introduced observational learning as the link for cultural transmission of behavioural traits and how canids learn behaviours from conspecifics. Then in Chapter 5, I outlined the importance of scent post maintenance for territory inheritance. The breeding season is the most active time for dingoes because they are either searching for a mate, a new territory or are working overtime to reduce the threat of invasion by extraterritorial foragers. Other dingoes outfitted with GPS collars that stored one location every 10 minutes showed similar patterns in movement where they walked to their border, and returned to their core area frequently. Movement and activity patterns, however, will differ invariably per site. Scientists in central Australia, for instance, have observed dingoes using roads and rubbish dumps as areas commonly traversed in search of food.

Figure 6.8 shows a conceptual framework of patterns of dingo movement, showing how they may forage and patrol borders within their territory and territory interstices, and how dingoes travelling between territories utilise the interstices. The dark shaded areas represent the highest frequency of scent marks of a pack (the 50% core area) and borders of each oval represent the home range, or the outer border of their territory. The non-shaded areas where their borders overlap represents the interstices of territories, which generally include shared resources. The interstices also are movement corridors for dispersing dingoes, or extraterritorial foragers like *Daoure* and are most likely to consist of ridges and tributaries. The movements of dingoes inside their home range are represented by thin black or thick white arrows. These are illustrating territory maintenance behaviours where pack members refresh scent posts to signify to intruders or neighbours that the territory is occupied and is being maintained. The real question here is what happens when territory maintenance stops?

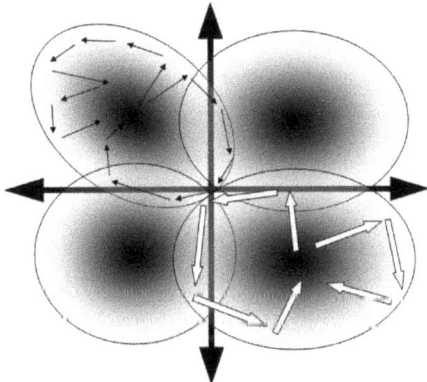

Figure 6.8 Conceptual framework for four neighbouring dingo territories, showing movement patterns of pack members within territories (→;⇨) and movements of extraterritorial foragers or dispersing dingoes between territories (➡). Note that core areas (shading) do not overlap in this diagram, but in wild situations, core areas of lone dingoes may overlap with the core areas of packs. Interstices of territories are depicted by the overlapping home range areas (oval borders). The core areas do not overlap because the frequency of scent marks increases to the point that intruders are aware that the area will be defended.

Patterns of activity

Activity is generally defined as the rate of movement between successive locations. Activity patterns may be classed as: a) resting (including eating) – when dingoes move less than 199 metres per hour (m/hr); b) walking – when dingoes move between 200–1999 m/hr; c) trotting – movements between 2000–5999 m/hr; and d) running – speeds greater than 6000 m/hr. Categories of activity may vary per location, however, depending on the habitat types and available resources. After defining categories of activity, movements may subsequently be calculated every five minutes, every 10 minutes, every 30 minutes or every hour; however, it is best to use the shortest period between points as possible.

Dingoes in Western Australia spent the majority of their active parts of the day travelling (49%), or performing other activities (46.5%). When dingoes were active, the least amount of time was spent hunting (4.5%). On average dingoes moved 3.3 km between locations and lone dingoes travelled further on average. One dingo during a 10 hour nocturnal period returned to its starting point, after walking 19.6 km. Dingoes in the north-east tablelands of NSW, however, showed individual variations in activity but generally had activity peaks in mornings and evenings. On average dingoes in the NSW north-east tablelands moved between 0.8 km and 1.2 km between consecutive fixes; however, the longest continuous distance moved was 9.8 km over five hours.

March and April were the most active months for dingoes in the Blue Mountains. Following the peak in activity, May was generally the month

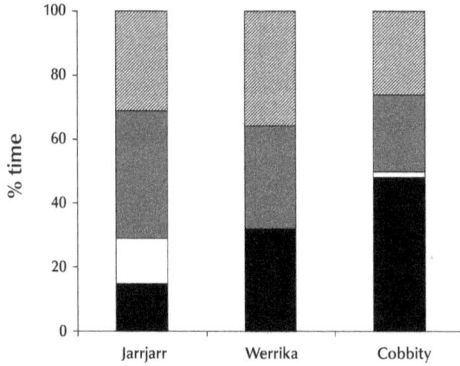

Figure 6.9 Percentage of time spent resting (black), in low activity (white), in crepuscular activity (dark grey) and in increased (diurnal and nocturnal, short- and long-distance) activity (light grey) for *Jarrjarr*, *Werrika* and *Cobbity* who lived in adjacent territories.

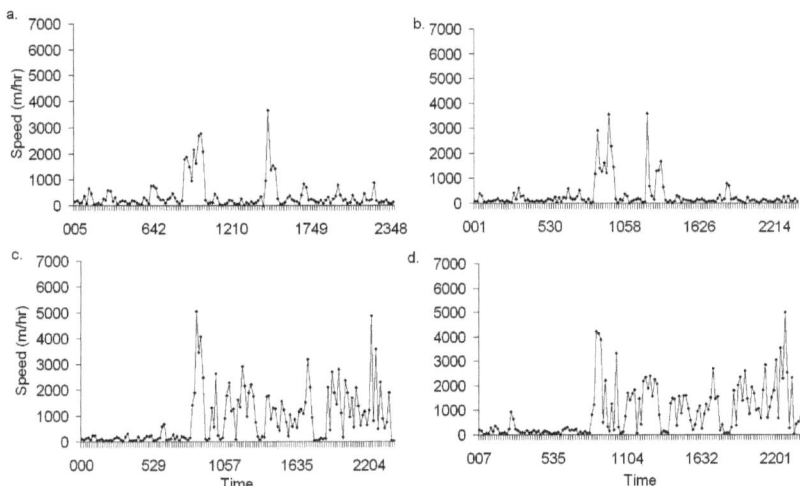

Figure 6.10 Levels of activity for *Willinga* (a) and *Jarrjarr* (b) on a normal day and levels of activity for *Willinga* (c) and *Jarrjarr* (d) during border patrol movements.

with least movement and movement slowly increased until February when it became reduced before the peak in March. Daily dingo activity was generally crepuscular during the 2007 breeding season and activity usually began at 6 am and finished by 8 pm with less activity between 10 am and 2 pm.

There were substantial variations in mean patterns of activity for *Jarrjarr*, *Werrika* (goanna) and *Cobbity* (white clay), three males that lived in adjacent packs (Figure 6.9). *Cobbity* travelled extraterritorially, *Werrika* once spent nine consecutive days resting for more than 70% of the day and *Jarrjarr* was observed circumnavigating his 95% home range.

Figure 6.10 is an example of the difference between an average crepuscular activity pattern and a border patrol activity pattern. The crepuscular pattern perfectly reflects a 'dog's life', and they probably only rose from their rest to eat, drink and play. During the border patrol, however, *Jarrjarr* and *Willinga* travelled approximately 17 km on the first day and 11.2 km on the second day. Their average speed was approximately 810 m/hr.

Combining techniques

The locational data collected using GPS telemetry enabled quantification of short-term daily movements, medium-term seasonal movements and long-term annual movement patterns with great precision. Other dingo researchers around Australia are currently using similar systems and I expect that, over the next 10 years, scientists and land managers will have an extensive inventory of dingo movement patterns. Research projects in the last few years have commenced in central Australia; South Australia; the Victorian Highlands; the Kosciuszko region and again in mid-coastal NSW; northern, western and urban south-east Queensland; and once again in Western Australia. Combined, these research projects will show effects of prevailing environmental circumstances and anthropogenic disturbances to dingoes, and of dingoes to humans.

Using GPS collars and motion-sensing cameras in the Blue Mountains, we observed some behavioural traits of communal living in hypercarnivorous canid populations. Pack territories of dingoes were consistent with the locations genetically related individuals were trapped. Sociality was observed and ranges were distinctly associated with prominent landscape features and movement corridors for extraterritorial foraging or dispersal.

One obvious movement pattern not yet discussed is that genetically related dingoes were trapped around the same time, and in the same area. The succession of captures was a pair from one genetic group, then another three pairs from different genetic groups on different nights, followed by five individuals from the same genetic group that also were related to the first pair captured, and then four individuals that were related to the second pair captured. This most likely indicates how packs moved into the interstices of their territories (where the traps were set) consecutively every two to three days, perhaps even after trapped and released dingoes had retreated to protective cover, or their core area. In addition, using the genetic data, GPS data, and images from motion-sensing cameras, we observed that a new pack formed during the study period. This pack was a derivative of the Douglas Scarp pack and dispersion of their genes through the Blue Mountains may increase variation in colouration. Conducting further analyses on the effects of movements by dingo groups on each other will be useful for management, especially if culling is the method of 'managing' a population.

Simple presentation of GPS locations has the ability to depict travel routes, rendezvous sites, spatial separation and overlap of territories, interactions between individuals and more importantly, identification of core areas. GPS data clearly display crucial differences between movements within core areas, territory maintenance (such as border patrols) and exploratory behaviour. Home range terminology, therefore, needs to be redefined to suit changes in technology as they occur. If exploratory activity, where dingoes travel between or through other dingo territories and return to their home range, were to be considered a part of a home range then it might be better termed as a dispersed range. A 50% kernel can remain defined as the core area of activity and the 95% kernel can remain defined as the home range. The dispersed range of *Daoure* observed during the 2006 breeding season, or the 100% MCP area traditionally deemed to be the home range, was 95.6% larger than his core area. He didn't even visit every square metre while on his foray either, but instead travelled at an almost continuous pace along roads, rivers and ridges.

Another important factor depicted from GPS data was the identification of core areas. The relationship of individuals and, in some instances, packs with their core area indicated core areas are a highly valued part of a territory. Studies on territory defence and inter-pack interactions in Ethiopian wolves suggested that intruding Ethiopian wolves identified territories by the frequency of scent marks. Strong site fidelity of dingoes to their core areas would suggest that the frequency of scent marks or strength of scent must increase to a point where it is beyond doubt to the intruder(s) that they have entered the territory of a pack. The only core area which overlapped with a core area of a neighbouring group was from an omega or lone female, so core areas that overlap may be representative of a range for a lone dingo. Social organisation of the animals under study has to be taken into consideration when attempting to interpret home range size and space utilisation by communal-living canids. Different dingoes may have different roles in the pack, different sized territories and different patterns of movement.

Reasons for the selection of a core area by a pack are subject to debate in the Blue Mountains area because there are no known limits in resources. In Western Australia, food availability was the determinant of the size of a core area for dingoes. This is consistent with the hypothesis that home range size is related to resource abundance within the optimal space. If food, water and habitat were homogenous and maximum pack size or threshold was achieved, then the home range and movements of dingoes is only limited by intraspecific competition or variations of social systems. Observations of extraterritorial foraging and mortality of males and females showed traits similar to other findings on the tolerance of conspecifics and suggested that tolerance of conspecifics was low during periods of breeding and rearing.

Of the 47 captured dingoes in the Blue Mountains, the proportions of males and females potentially dispersing were both 13%. A higher proportion of females (13%) forayed extraterritorially compared with males (10%). Of the 47 captured dingoes, nine subsequently died, eight travelled extraterritorially, and six may have dispersed. Seasonal and monthly observations showed all females that died or dispersed did so at the end of whelping or in the rearing season and males generally had two peaks, one during the breeding season and another during whelping season. In contrast, sightings or camera observations of groups were lowest during whelping season (five sightings) and increased during rearing (seven sightings), exploratory (10 sightings) and eventually peaked during breeding season (14 sightings). Dispersal and mortality data of captured females showed that four of the five observations were from genetically related females in the same area. In addition, one of the collected skulls from a female had a hole the size and shape of a canine tooth in the top, which implied intraspecific predation may have been the cause.

Effect of social systems on movement patterns

If the dominant breeders in a dingo pack want their next fittest progeny to inherit their territory then they need to manage their pack to account for variations in food supply and changes in the composition of neighbouring packs. They therefore need to defend their territory until they weaken or die and the next fittest progeny can take the reigns. Any observed intolerance of some pack members by the dominant breeders is an indication that the dingo pack is stable, functional, and subjected to variations based on resource abundance and environmental conditions. Dingo societies will always remain dynamic because if they live in habitat with abundant resources then the packs can increase in size until they have a negative effect on resource abundance. Then in situations where resources have become less abundant, groups are likely to fragment and reform to a suitably sized population level.

Overlap of territories indicates that interstices of territories and any associated resources are habitually shared between neighbouring dingo packs. Thus when one pack leaves the shared area, another can enter. I made observations of this on numerous occasions using motion-sensing cameras. At the start of some three-day observation periods I would either photograph or would see the Campbell's Creek pack or the Douglas Scarp pack, and on the third day it would be the opposite pack to the first. These observations are consistent with interactions between wolf packs sharing moose herds and dingo packs sharing water resources in central Australia.

Variations in dingo society each season also clearly affect dingo movement patterns. The GPS movement data from the Blue Mountains showed that modern day dingoes remain primitive hypercarnivorous canids characterised by one annual breeding cycle, as defined at the end of Chapter 3. Scientists that studied feral dogs in central Italy showed that feral dogs continued to breed and attempted to rear young biannually in a resource-saturated site (they lived between two garbage tips). They also showed that the feral dogs did not interact or breed with wolves because the wolves avoided human settlements. All movement patterns for males and females in the Blue Mountains showed core areas reduced in size during whelping and slowly increased in size as the pups aged. Speed and distance travelled between points slightly increased each month while pups presumably developed their skills, until the next whelping season when the dingoes moved much less. Breeding trends observed in both sites remained annual, opposed to biannual as has been suggested as a potential threat from 'hybrid dingoes' or 'wild dogs'. People have been fearful of this occurring due to the introgression of domestic dog genes. Dingoes have, however, been observed to raise two litters from each of two females within one pack in Western Australia, and one wolf pack also reportedly raised two litters borne by different mothers. The Blue Mountains is a resource-abundant site and there was no evidence that Blue Mountains dingoes can either raise two litters within one year or that females go into oestrus twice. Reproductive suppression by dominant animals in hypercarnivorous canid populations is yet another factor that can influence the movements of individuals. This is especially so for unwanted sub-adult females that probably only have the options of fighting or defending themselves until they are killed or take their chances through emigration. Studies of movement patterns can therefore be used to demonstrate the functional role of dingoes by showing their discrete demarcated territories and seasonal patterns of extraterritorial explorations.

The family Canidae are landscape specialists. Wolf biologists have reported observations of wolf movements that suggested wolves have complex memories of landscape features. In all studies of dingo movement patterns, researchers showed that they utilised ridges, gullies and other prominent landscape features when foraging and patrolling borders. Data from the Blue Mountains that showed the dingoes' maintained peripheral borders encompassing their core area all year round indicated that territories are constantly marked and defended. This was also similar to results reported in wolf studies that showed one nursing female wolf continued to patrol the boundary of her territory in early stages of rearing. All females in the Blue Mountains maintained movement patterns to home range borders in every season. Maintenance of scent posts may be necessary for every member in a pack of dingoes to: a) signify that the pack remains functional; b) to protect their young from predation by competitors; and c) to maintain olfactory communication with neighbouring groups.

Studies on dingo activities and behaviours during border patrol movements should be a priority for future research to assist land managers because dingoes tend to be managed on the borders of their territories. If *Willinga* and *Jarrjarr* were killed during their border patrol, would their former territory be invaded by dingoes from adjacent packs? Could the remaining members of their pack defend their territory? If not, would remaining members disperse and attempt to become established elsewhere, as has been observed in Western Australia, or would they simply die? Research to improve dingo control has rarely accounted for the impact of negative effects on the sociobiology, nor has it considered the impact of dingo control on Australian ecosystems. Research assessing the effects of secondary poisoning or the effects of traps on non-target animals has been undertaken but these were generally in response to public concern regarding animal welfare issues or to review products for improved dingo control.

Robert Harden showed in his study in the NSW north-eastern tablelands that dingoes did not travel from the core of protected areas to kill livestock. The common theme for dingo migration from his observations and observations in other studies then is reproduction. Controlling oestrous cycles of domestic dogs or dingoes living adjacent to protected areas could reduce dingo movements around the periphery of a park when dingoes are actively seeking a mate. This would presumably reduce interactions between dingoes and livestock and minimise the effects of dispersal sinks, which are created when dingoes are controlled using the buffer zone technique. American scientists have recently developed a buffer zone with sterile coyotes to reduce the risk of hybridisation between coyotes and red wolves, so why not trial a buffer zone of sterile domestic dogs in Australia?

Female dingoes in the Blue Mountains were not observed outside their common range until the whelping and rearing seasons when their mortality rate also appeared to increase, coincidentally when pups begin to forage. This is the same time that the dominant females would show most aggression towards excess, resource-consuming females. Male dingoes alternatively were not observed outside their home range until the breeding season. Relating the data to annual dingo biological cycles showed that extraterritorial forays were motivated by instinct to increase opportunities to mate or potentially to escape death. *Daoure* travelled an approximate route greater than 120 km from the most eastern GPS locations of his extraterritorial forays, west, to property housing and breeding captive dingoes when the GPS collar logged approximately 17 locations within 3 km of the property. At the time, between 18 and 22 dingoes were thought to be housed on the property. If 7–10 of these dingoes were females in oestrus surrounded by 10–12 males howling from a captive location then it could be assumed many more wild dingoes within the Blue Mountains would probably be attracted to that property. Could dingoes be attracted to livestock enterprises because they are housing domestic dogs in oestrus without sufficient scent posts to deter intruding dingoes?

To maintain the integrity of future research on dingo movement patterns in areas adjacent to livestock enterprises, it would be beneficial to include dingoes living on park boundaries and in the core of protected areas, and domestic dogs from properties adjacent to dingo habitat. Movements of free-ranging domestic dogs have been seldom studied, yet domestic dogs are known to predate livestock and native fauna. Studies on feral dogs in central Italy showed that feral dogs were 'merely a nuisance' to land managers. On the contrary, free-ranging pet domestic dogs were reported to kill three calves in a study in America and domestic dogs have also been observed killing livestock in Italy.

Pup rearing was determined to be the most important seasonal factor influencing dingo movements in Western Australia. This was also observed in the Blue Mountains; however, territory maintenance was the overarching pattern of movement. Apart from spending 50–70% of their time resting, dingoes tend to spend the remainder of their time traversing their home range. If they aren't hunting or drinking, then they are most likely updating their scent posts on the edges of their core area or on the edges of their home range. I have watched dingoes playing after feeding on a kill, standing in water on a hot day in summer and lazing about by their den and by a river. Dingo movements in the NSW tablelands also were short and interspersed with rest periods, and they were neither nocturnal nor exhibited crepuscular patterns, which implied that they to were cathemeral (irregular). Although dingoes spend substantial amounts of time resting, activity can be markedly different on some days compared with others. On average one day in eight had more nocturnal activity than the others and other patterns in activity between days occurred irregularly. Due to these variable activity patterns, it can be speculated that external factors such as weather, moonlight and temperature may be influencing dingo movement patterns. Observations of activity may also be affected by the sampling regime set by the researchers. Scientists in Queensland did not examine daytime activity because dingoes had previously been reported as nocturnal or crepuscular, but there is more evidence that suggests dingo activity varies per site. Patterns of activity in the Blue Mountains showed dingoes were more diurnal than nocturnal, that activity patterns across consecutive days are not predictable and that activity patterns varied per dingo. Similar observations have been made for gray wolves and for maned wolves, *Chrysocyon brachyurus*.

How many dingoes are there?

Extrapolating data from the size of observed packs and the size of their home range in the Blue Mountains, and knowing that my study area was approximately 220 000 ha enabled me to estimate how many individuals and packs lived within the area. If each pack within a mean home range of 37.7 km^2 consisted of five dingoes (aged >1.5 years) and four pups during whelping season, the estimated population for my study area was approximately 467 dingoes distributed through 58–59 packs annually. This model has inherent flaws because it does not and cannot account for such factors as social cohesion of groups, survival estimates, habitat variation (dingo home range in more rugged terrain was generally double the size of those in ex-pastoral lease habitat) and selective management of the population across this vast area. If 61.8 km^2 (average home range size in rugged terrain) was considered the mean home range of dingoes for the entire southern Blue Mountains, then approximately 285 dingoes are distributed through 34–35 packs. An average of the two estimates is calculated as 376 dingoes, distributed through 46–47 packs. Understanding now that dingoes are governed by an intricate social network that exists within and between packs, then the next question on the tip of my tongue is how do dingoes and prey interact?

7

THE ROLE OF A HYPERCARNIVOROUS PREDATOR

Dingoes are often referred to as a *keystone species*. A keystone species is a species that exerts a large, stabilising influence throughout an ecological community, despite its relatively lower population size in comparison with other species present. Removing a keystone species from the landscape is said to have strong effects on diversity in animal communities and the composition of animals in those communities. In the western NSW region, where dingoes were heavily controlled and, to a degree, exterminated from within the dingo fence, 24 native mammal species have since become extinct. Some researchers suggested that feral animals and the rabbit plague were the most significant contributors to this extinction crisis. Ecologist Dan Lunney from the NSW Department of Environment, Climate Change and Water stated that the extinction process was largely attributed to sheep farming and sheep farming processes. That includes removing a keystone species.

Dingoes have direct effects on the daily business of nature. Their presence in some areas will cause prey species to reside elsewhere, and their eating habits may reduce the abundance or composition of the prey species assemblage (the taxonomic subset of a community). In addition to predator–prey interactions, however, are predator–predator interactions. One theory known as mesopredator release, questions the effect hypercarnivorous predators like the dingo or the gray wolf, have on mesocarnivorous predators like the red fox or the coyote respectively. So, in the example from western NSW where dingoes were in very low densities for approximately 100 years, there were no interactions between hypercarnivorous predators and prey species that weigh more than 15 kg, nor was there any competition for foxes and cats. Therefore, extensive landscape modifications by humans enabled the large herbivores (kangaroos and domestic livestock) to exploit the landscape, and the mesocarnivores (foxes and cats) to exploit the small mammals. That is the recipe for Australia's extinction rate, the highest in the world.

To understand the role of a hypercarnivorous predator, or a keystone species, we need to investigate how dingo presence affects the presence of prey species and competing species. This can be done through comparative analyses of diet and abundance. The diet of dingoes is a relatively common subject and tends to be investigated concurrently with dingo movement patterns. This is most probably because showing the occurrence of livestock as a food item with the movements of dingoes on or through livestock enterprises is the easiest method to justify dingo control. Another reason is because some people think that having an understanding of dingo predation can maximise livestock production. Past research on diet, however, consistently showed dingoes cared much less about livestock than they did about native prey. In addition, hard evidence that dingoes emigrate to livestock enterprises for lamb, mutton or beef are more than scarce – they are absent. Most researchers find instead that emigration occurs potentially when individuals are seeking a mate, to escape conflict with conspecifics, or a combination of escaping conflict, finding a mate, and resource availability.

How abundant or how active dingoes are in an area are two further research questions commonly asked. Activity in this form is different to the rate of movement between successive locations and instead measures either the proportion or the frequency of dingo tracks on fire trails during set periods of time. Usually the set periods of time are before and after control campaigns, to measure effects that the campaign may have had on the population. These measures can also be used to investigate the abundance and activity of prey species.

If we want to understand what role Australia's hypercarnivorous predator might play in Australian landscapes, then we simply have to correlate dingo eating habits and dingo abundance/activity with changes in abundance/activity of prey species and competing species. For instance, if the activity of dingoes increases simultaneously with the activity of kangaroos, and kangaroos increase as a food item as a result of increased dingo activity or increased kangaroo activity, then we have seen a direct predator–prey interaction. It is unknown if the kangaroos increased their activity because they were being hunted, or if dingoes ate more kangaroos because kangaroos were more abundant or active.

Developing an understanding of these interactions helps us move one step closer to answering the one question on everybody's lips: are dingoes Australia's fabled top-order predator? Originally, it occurred to me that the most complex part of a dingo was how it communicated with other dingoes. Don't get me wrong; the social structure of a high density, closely related dingo population living in optimal conditions is pretty hard to grasp when you first start seeing the data unveil itself. But these predator–prey and predator–predator interactions complete the puzzle and show us why we need to understand dingoes so intimately. To put it succinctly, the functional role of Australia's top-order predator is to be a top-order predator.

As was stated in Chapter 4, basic data on seasonal changes in abundance and activity of dingoes are scarce. In addition, data on abundance, activity and even diet are often constrained by logistical limitations; however, extensive datasets need to be compiled to gain an understanding of what effects dingo abundance or activity have on dingo diet, and on prey abundance and activity. Past interactions observed between dingoes and their prey showed diet of dingoes alternated with fluctuations in prey abundance. Prey switching, as it was called, was relative to changes in prey abundance and shifts in weather patterns in central Australia. To attain these data on prey switching, scientists analysed the contents of dingo stomachs to determine interactions between predators and prey. Once again, if we return to the sociality of dingoes, then culling dingoes will have detrimental effects on their ability to function as a communal-living, top-order predator.

Studies on dingo ecology have begun to focus on the role of the dingo as a higher order predator. Some researchers have suggested that dingoes may aid the conservation of Critical Weight Range fauna (CWR; 500–1500 g) by suppressing the abundance and activity of mesocarnivores (cats and foxes) which prey opportunistically on species that weigh less than themselves. Comparative tests in various studies of predator–predator and predator–prey interactions around Australia have since provided evidence that dingoes may influence movements and foraging effort of competitors and prey.

Variation in dingo diet

Since dingoes range over a much wider area than any of their prey, different populations of dingoes predate different populations of prey. In each region, one or two prey species tend to be the major sources of food. Most studies in eastern and south-eastern Australia have shown that swamp wallabies, wombats, eastern grey kangaroos and brushtail possums were the dominant prey of dingoes. However, in Western Australia euros and red kangaroos were dominant food items, and in central and north-central Australia, rabbits, reptiles, agile wallabies *Macropus agilis* and magpie geese *Anseranas semipalmata* were dominant food items.

Researchers have stated that the abundance and status of dingoes has changed since European settlement and as a result, so has the nature of dingo predation. Following discussion of habitat fragmentation; overgrazing and increased competition by rabbits and livestock; altered fire regimes; increased water availability from artesian bores; and the effects of introduced feral cats and foxes, *dingoes* were suggested to be responsible for decline of some native marsupials and extinction of others! These scientists also stated that the predatory cycle alternates between consistently available and seasonal prey.

What then does consistently available and seasonal prey constitute? Flush and drought periods in central Australia were referred to as times when predators and prey took advantage of arising opportunities. In one instance, rabbit populations increased when dingoes were being controlled, and in another the rabbit populations appeared to be regulated by dingoes. Magpie geese were apparently most often eaten as fledglings, and rats were most often eaten when their populations increased, but geese and rats are floodplain fauna and only accessible in dry periods. In north-east NSW, dingo predation disrupted the seasonal pattern of swamp wallaby births because the females ejected their young so frequently to escape predation. Kangaroos are also known to eject their young when being harassed by dingoes and it has been hypothesised that this may affect macropod recruitment rates. Whether these prey species are consistently or seasonally available is yet to be determined. Magpie geese may migrate in the dry season but to my knowledge, no research exists to suggest that macropods, possums or rabbits migrate. Those prey species can increase in abundance at certain times of the year when they breed, but dingoes also may appear to increase in abundance when they are actively searching for mates or decrease in abundance when rearing pups and movements are consolidated. All of the common food items for dingoes in south-eastern Australia can breed throughout the year; however, most populations of brushtail possums have a major autumn and minor spring breeding season and studies on seasonality in swamp wallabies stated that it varies from site to site. In one site where dingoes were excluded, swamp wallaby births peaked in March and no young were born between August and September.

It is important to remember that dingo social systems will most probably regulate the use of resources within a territory. Any observed dispersal by dingoes from their home range, for instance, is due to the effect of social systems or lack thereof, and social systems are in place so resources are available for the pups of the dominant pair. You may recall from Chapter 4, that dingoes have four major biological seasons: breeding, whelping, rearing and training. In the breeding season, the age structure of the dingo population is highest and because it follows the training season, most dingoes should be adept hunters. Whelping and rearing seasons are obviously when pups are youngest and adult movements are somewhat restricted by rearing activities. Therefore, to understand dingo diet we need to correlate changes in the abundance and activity of dingoes with changes in the abundance and activity of their prey.

In our Blue Mountains study, we collected and analysed scats from dingo and fox scent posts to investigate diet on a monthly basis for 26 months. Simultaneously, we gathered data on activity and abundance of the predators and medium- to large-sized vertebrate prey. From a total of 2451 scat samples we identified 27 mammalian species, dingo samples contained 20 species and fox samples contained 22 species. Compared with dingoes, foxes appeared less selective of prey, with more variation, opportunism and more species from the critical weight range.

Outcomes from these data were diverse. In one test we used the Predator Diet Index (PDI) and showed that the proportions of prey species varied for dingoes and for foxes (Figure 7.1). In both forested and ex-pastoral lease habitat, swamp wallaby was consumed by dingoes mostly in their breeding and whelping seasons. Proportions of prey in dingo diet then switched, and consumption of brushtail possum increased during the whelping season and it was as common in dingo diet as swamp wallaby during the rearing season. In the ex-pastoral lease habitat, a similar trend was also observed for eastern grey kangaroo, and carcasses were often littered around dingo den sites. Perhaps the adults were dragging kangaroo carcasses back to the den for their pups. Foxes, however, showed no clear trends that suggested they switched between prey species or had seasonal preference of prey.

Comparison of dingo diet with fox diet clearly showed differences in prey selection and that dingo prey selection was most closely associated with dingo biological seasons. Differences between diet of a top-order hypercarnivorous predator and of an opportunistic mesocarnivorous predator were also apparent. The next question then is: was there any observed increase or decrease in prey proportions related to activity of prey or of prey biological seasons? Since we did not collect any data on reproductive biology of prey species we could only use speculation to identify reasons why prey activity may have increased, or why variations in proportions of prey species in diet occurred. In most cases, however, the activity or abundance indices of wallabies and possums were higher when their PDI was lower. In other cases, activity or abundance indices of prey fluctuated dramatically while the PDI

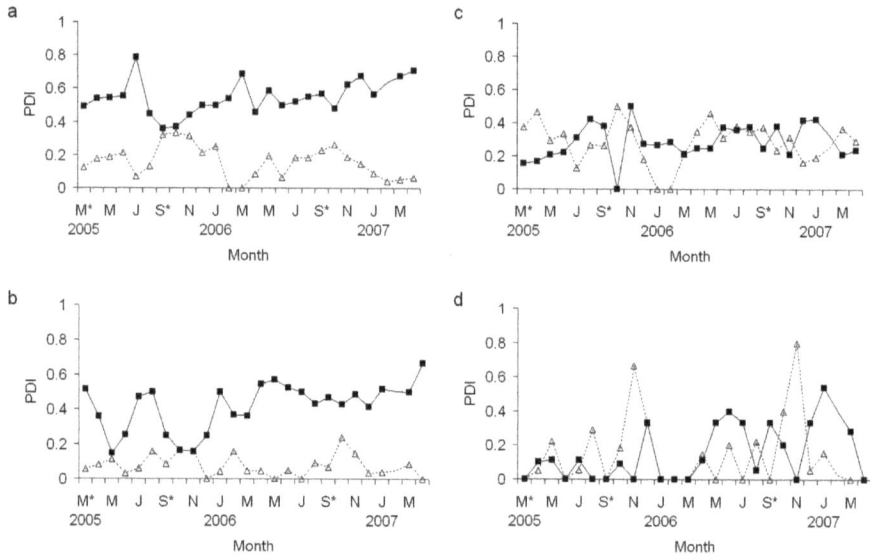

Figure 7.1 Proportions of swamp wallaby (■) and brushtail possums (Δ) consumed by dingoes in forested habitat (a) and ex-pastoral lease habitat (b), and foxes in forested (c) and ex-pastoral lease habitat (d).

showed the prey was being consistently consumed by dingoes.

Other studies have subliminally made the same observations. Dingo predation on red kangaroos increased during January and March in Sturt National Park, NSW. In total, 83 red kangaroos were killed by dingoes within 150 m of a water source in that study. Relating these data with dingo biological seasons indicated that the dingoes studied may have been training juveniles how to hunt kangaroos, especially because the predation rate decreased by March when the dingo breeding season commenced.

Differences in the composition of diet in north-east NSW were observed between May and October when non-macropod species, such as possums, increased relatively and November and April when predation of macropods increased. These observations are directly comparable with data from the Blue Mountains when brushtail possum increased in the diet of dingoes between June and November in 2005 and 2006 and swamp wallaby increased in the dingoes' diet between December and March in 2005 and 2006. Dingo diet therefore alternates during the annual reproductive cycle of dingoes and subsequent changes in the hunting strategy, and in the age structure of the dingo population.

Evidence from a seven-year study on wild dingo ecology in central Australia resulted in the concept of 'alternation of predation'. The hypothesis was that dingoes alternated between large and small- to medium-sized prey during periods of drought and flush periods respectively. It was proposed that this interaction with prey and the environment was consistent with all models of foraging behaviours. One major difference in that study compared with our project in the Blue Mountains, is that diet was determined from contents of 386 stomachs. Dingoes from central Australia were culled to collect samples, whereas we used less-invasive techniques. Removing dingo and fox scats may have confused some individuals into thinking that they had missed a scent post during a border patrol, but they at least had the opportunity to re-scent each post and resume normal hunting activities.

Initially in central Australia, 58 stomachs were sampled during the first sampling period, followed by 49 in the second. Later sample sizes remained below 35 dingo stomachs per sampling period with the exception of one year, when 42 stomachs were collected after the false break in a drought (sample sizes: 58, 49, 18, 31, 23, 34, 35, 30, 35, 42, 18 and 13 over seven years). This sampling technique, which successively removed dingoes from the population, would have changed the age and pack structure of dingoes within their population following repeated regular sampling. The ability of some individuals to hunt in pack formation and hunt large prey may also have been affected. Perturbations in dingo social systems have been suggested as a mechanism that enabled solitary dingoes to establish a territory in areas no longer defended by packs, which also indicates neighbouring pack ranges could expand. Researchers in Queensland and South Australia have also showed that predation on livestock increases following dingo control campaigns that may have disrupted the dingo social system. Scientists conducting the study in central Australia were possibly, therefore sampling diet of dingoes from an agitated population. Dingo abundance in that study appeared to have been reduced after the first two sample periods since numbers of stomachs sampled, or numbers of dingoes available to be shot or trapped, did not approach the same sample size for four years. Proceeding the fourth year when 42 samples were collected, dingo numbers had obviously decreased, having two of the three lowest sampling events in seven years. It is most plausible to assume that results from that study showed variations in the feeding ecology of dingoes during seven years of high, human-induced dingo mortality during climate-induced fluctuations in prey populations.

Variation in dingo abundance/activity

There are a number of techniques available to investigate dingo abundance and activity. Modelling densities based on home range size and pack size and data from motion-sensing cameras are two novel techniques, while counting tracks on sand plots and scats per kilometre are common traditional techniques. Number of scats collected on fire trails, per kilometre, per day, per site and month was used to investigate changes in relative abundance of dingoes in the north-east NSW tablelands. Scats generally increased in abundance from June through to November. To develop similar indices in the Blue Mountains, we compared the frequency of dingo tracks on sand plots with the number of scats collected per kilometre each month (Figure 7.2). Trends indicated that faecal abundance increased when dingo activity decreased and faecal abundance decreased when dingo activity increased. These data reflect the movement data observed in Chapter 6, where the home range size remained stable while the core area size and activity decreased during the whelping and rearing seasons. These trends also indicate that:

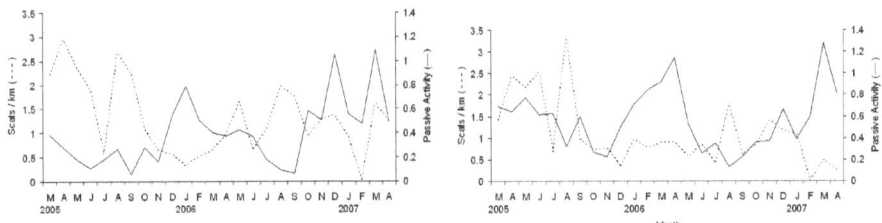

Figure 7.2 Comparison of scats collected per kilometre (- -) on road infrastructure with the Passive Activity (—) of dingoes on road infrastructure in forested (left) and ex-pastoral lease (right) habitat in the Blue Mountains.

- dingoes may use scats when there are fewer individuals marking territory with urine;
- scats deliver a different message about the activities of territory holders to conspecifics; and
- scent post maintenance is a regular pattern of movement.

Data on activity and abundance of dingoes in the Blue Mountains showed that levels of dingo abundance or activity naturally fluctuated, without being subject to control campaigns. This throws doubt on the use of sand plots to determine the success of control campaigns. Interpretation of changes in dingo abundance and changes in dingo diet in coastal south-east Australia around Nadgee Nature Reserve showed various annual trends over a nine-year period. Due to a wildfire that happened in the second year of the study, researchers investigated restructuring of mammal communities and interactions between dingoes and prey. It can be speculated that measures of dingo abundance may have increased because scent posts would have been consumed by the fire and packs would need to remark their territories. The dingo population may have increased to aid territory defence though it is more likely that the levels of dingo activity increased while territories and packs were reforming, especially when abundance tests show similar trends to activity (Figure 7.3). Predation of macropods also increased post-fire when dingo abundance increased. Pending reformation of dingo social systems and the kangaroo population, one may hypothesise that macropod predation would eventually revert to seasonal patterns. Alternatively, the study may have showed that dingoes in Nadgee Nature Reserve maintained the equilibrium of the ecosystem by suppressing populations of herbivores during instability in the environment.

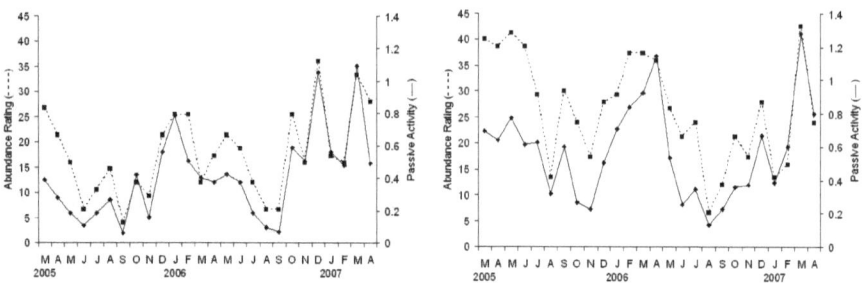

Figure 7.3 Comparison of relative abundance rating indices with passive activity indices per month in site 1 (left) and site 2 (right) for 26 months in the Blue Mountains.

Relationship between diet and activity

Whenever a prey species, such as a kangaroo, wallaby, wombat or possum or even a lizard sees a predator such as a person or a dingo, they run and hide or watch intently in case they have to scramble to safety. That is one form of interaction between predators and prey and it is as simple as saying predators make prey run. Scientists often use mathematical tests to assist their understanding of the significance of prey running from predators, or predators running after prey. The tests essentially show whether or not the prey species in question is eaten by the predator when the predator is more active or less active, or whether the prey species are more or less active when the predator is more active or less active. Scientists can then state whether there was or was not a 'co-relationship', or more technically a correlation, between the predators and the prey. If an increase in prey activity correlates with an increase in predator activity, then there is a close similarity or connection between them, often because the predator is responding to the prey or the prey is responding to the predator. For instance, an increase in dingo activity significantly correlated with an increase in the activity of brushtail possums, wombats and kangaroos in the Blue Mountains. These data indicated that those prey were more active when dingoes were more active. There were various relationships, however, between predators, between predators and prey, and between different prey species. In some instances, activity of one prey species correlated with the occurrence of other prey species as dietary items for dingoes. Wallaby activity correlated strongly with occurrence of swamp wallaby as a prey item for dingoes and for foxes. Then there was a negative correlation between the occurrence of swamp wallabies as a dietary item for dingoes, and the activity of kangaroos. This suggested kangaroo activity was lowest when dingoes were consuming swamp wallabies or alternatively, dingoes changed to consuming swamp wallabies when kangaroos were less active. Occurrence of swamp wallaby in dingo scats positively correlated with occurrence of wombat, rabbit and brushtail possum in dingo scats.

In essence, the data showed that when kangaroos were more active, dingoes were more active and the kangaroos were being eaten by the dingoes. When kangaroos were less active, dingoes also were less active, but wallabies were more active. When wallabies were more active, swamp wallabies were being eaten by dingoes. These interactions are graphed in Figure 7.4. Columns above zero identify the animals that occurred more frequently as a dingo food item when the activity of the animal named on the bottom axis increased. Columns below zero identify the animals that occurred more frequently as a dingo food item when the activity of the animal named on the bottom axis decreased.

These data show the complexity of predator–prey interactions and possibly show the functional role of dingoes at a landscape scale. Wallaby activity was associated with the occurrence of swamp wallaby as a prey item for dingoes and foxes. The negative correlation between occurrence of swamp wallaby in dingo diet with kangaroo activity on roads suggested that decreased activity of kangaroos occurred simultaneously with increased foraging efforts by dingoes on swamp wallabies or vice versa. Kangaroo activity was higher when kangaroo increased as an item of diet when dingo activity on roads increased. These data suggested that dingoes may influence

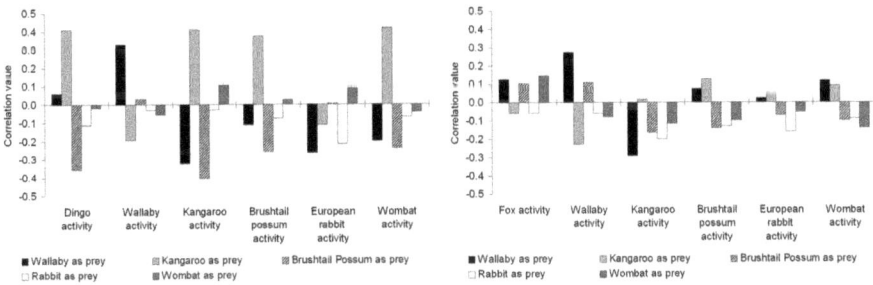

Figure 7.4 Relating dingo (left) and fox (right) activity with prey activity and the occurrence of swamp wallaby, eastern grey kangaroo, brushtail possum, European rabbit and common wombat as dietary items.

kangaroo movement patterns, or kangaroo abundance and activity influences dingo activity. Shifts in activity, abundance and diet of a hypercarnivorous predator may therefore affect activity and abundance of competing predators and their prey.

The role of a top-order predator

From the combination of old and new data sets we can state that:

- dingoes are selective of different prey species during their different biological seasons;
- dingoes possibly use scats to mark territories when pups are young, and use urine to mark territories when pups participate in territory maintenance behaviours;
- foraging behaviours of dingoes may provide periods for vegetation to regenerate in some habitats when dingoes shift predation from kangaroos or brushtail possums to swamp wallabies and vice versa;
- feeding ecology of dingoes changed with the reproductive needs or age structure of the dingo population suggesting adult dingoes train juvenile dingoes how to hunt and survive between whelping and breeding seasons, on an annual basis; and
- based on the combined evidence, it appears most likely that prey species react to changes in the activity of dingoes, and dingoes select prey suitable for the age structure or the cohesiveness of the pack.

Essentially dingoes may facilitate ecosystem processes through natural, repeated predatory behaviours during the identified biological seasons of the dingo; however, there is much more that needs to be studied and understood. Research on the additional functional effects of dingoes on competitors using the theory of mesopredator release is suggested to be a beacon for the conservation of threatened native species. For instance, researchers have shown that dingoes may effect fox populations in some areas and cat populations in other areas. Populations of other introduced pests such as feral pigs *Sus scrofa* and feral goats *Capra hircus* may also be affected by dingo predation. In the wet dry-tropics, predation of pigs increased simultaneously with growth in the pig population; however, the correlation was negative and data indicated that pigs may have also been affected by interference competition with feral water buffalo *Bubalus bubalis*.

On the selection of a core area of activity, dingoes appear to inhabit areas that provide optimal resources (food, water, protective habitat) sufficient to maintain daily energy requirements. Dingoes may, therefore, concentrate their activities towards territory maintenance for inheritance. Dispersing dingoes are essential for reducing genetic drift; repopulation of areas void of dingoes or areas affected by baiting/control campaigns; to become helpers for packs that suffered poor reproduction rates; or to assist packs that are suffering from high predation of offspring. Maintenance of the dominant pair to control order in their pack, which consequently maintains order between the surrounding packs within a protected area, ensures resources are not over exploited in core areas and territories. Evidently, dingoes are self-regulated dependent on resource availability and may not need to be managed or controlled as drastically as they currently are in some cases. They will still need to be monitored consistently so livestock producers can adaptively manage their livestock.

To understand the role of the dingo more fully in the conservation of Australian ecosystems, a holistic, landscape-scale analysis is required because canid species are adaptable to a variety of ecosystems. The role of dingoes may, therefore, be broader than the conservation of an ecosystem; they have a conservation role throughout the entire landscape. Due to this, there may be considerable variation in genetic population structure, hunting behaviours, activity patterns, and patterns in prey consumption by dingoes across Australia.

8

COMPETITION BETWEEN HUMANS AND DINGOES

Firestick farming by Aborigines is grounded in Australia's history and it served a number of purposes. These included a reduced risk of wildfires and modifying the landscape for travelling, or to create better habitat for hunting. Increased fire frequencies affect biota by removing the shrub layer and promoting growth in the grass layer. Kangaroos and wallabies much prefer to forage grasses in open habitat that is adjacent to closed habitat because they have better peripheral vision to watch for predators and an area to escape a predatory attack. Livestock enterprises are perfect habitat for kangaroos and wallabies because they are modified in a similar sense to areas that were traditionally farmed with fire, and they contain numerous water resources. Dingo control adjacent to livestock enterprises constantly increases availability of optimal habitat for dispersing dingoes because there are no dingoes left to compete with, and prey are abundant.

Dingoes versus livestock

In all dietary studies, livestock was a *minor* contribution to diet and predation of livestock or scavenging of carcasses could not be distinguished in most. Plus it was not known whether the livestock were feral themselves or if they were sick and dying regardless of dingo predation. Feral cattle were present in the Blue Mountains during my study and park rangers informed me of instances where dingoes crossed paths with cattle and the cattle showed no fear of predation. Livestock also may sometimes be killed by dingoes without being eaten. Research has not yet looked at why such 'surplus killing' events occur but there are a number of situations that the behaviour may be linked to:

1 juvenile dingoes being trained by adult dingoes how to attack that prey type;
2 the flight behaviour of the prey excites the attacking dingoes much like a domestic dog is excited by chasing a ball or working dogs are excited by running livestock;
3 dingoes that survived a control program attack livestock because they are in abundance and easy to kill; or
4 working dogs and feral domestic dogs that go wandering at night are responsible for the attack(s).

Research has shown that domestic dogs in a coastal town of southern NSW wandered as far as 30 km and returned to their neighbourhood in one day. This is consistent with point 4 above because the domestic dogs also congregated as a pack. Other research in America has documented livestock predation by domestic dogs, and a population of feral dogs in Italy was sustained by recruitment of stray dogs.

A controversial addition to most dingo studies is that some were directly questioning the impact of dingo predation on livestock enterprises and were funded by the Australian Meat Research Committee (Australian Meat and Livestock Corporation), Agriculture Protection Board of Western Australia and the Rural Credits Development Fund. Since research projects were funded to investigate the ecology and biology of dingoes, and to observe the effects of control techniques, then it is no wonder that the recommendations of past studies were to control dingo populations when they are most active or in higher abundance. Recently developed views hold that disturbing dingo pack structure, a consequence of dingo control, forces individuals to hunt alone and subsequently predate opportunistically on easy and abundant prey such as livestock. Research on cattle stations in Queensland showed that the highest levels of calf predation by dingoes occurred on baited sites opposed to non-baited sites. In contrast to the increased levels of predation observed, estimates of dingo activity showed that activity of dingoes, measured by the amount of dingo tracks counted on a transect such as a road, was reduced as a result of the baiting campaign.

'There is anecdotal evidence ...'

In Chapter 3, the question of what a dingo was, is, or can be in the eyes of the beholder was discussed. That section started with a myth by a livestock farming community and reported by popular media that dingoes, the dingoes specifically from our Blue Mountains study site, were migrating out of the protected area to kill their livestock. Roland Breckwoldt also raised this issue in his book, and quoted a dingo trapper in north-east NSW that claimed dingoes travelled extensively between sheep farms and 'hit' sheep farmers hardest in April, May (or early June), October and November. Robert Harden's research in that area showed no evidence for these claims of migration. In the life and times of a dingo, however, April/May is breeding season when dingoes look for a mate, and October/November is rearing season, the time of year that training begins for naïve dingo pups and excess female dingoes are, in theory, driven out of the pack. Conversely, the Australian Sheep Industry Cooperative Research Centre released a pamphlet in 2006 that showed the optimum time of lambing for sheep enterprises was related to pasture growth, and pasture growth rate was highest in the dingo-rearing season! The research on dingoes that was funded by livestock industries showed that this was the time when dingoes were rearing young and increased in abundance and movements. The reality of the situation is that dingoes have rarely been studied to reach a larger understanding of our place in the Australian environment. Dingoes have mainly been studied so humans can maximise the efficacy of control efforts, capitalise on available resources and increase short-term economic gain.

There are a few important points associated with economics that need to be made here, and they are interrelated. Firstly to set the scene, appropriate management for conservation of endemic carnivore species is of concern to ecologists because of their functional role and importance in terrestrial ecosystems on which *all* life depend in one way or another. Native carnivores may shape the demographics of prey populations; suppress populations of introduced species that have deleterious effects on native species, ecosystems and landscapes; attract tourists and encourage conservation of protected areas; and generate income and provide employment opportunities for rural communities. Carnivores, however, tend to prey on species valued by humans and any available employment opportunities may involve the suppression of carnivores – Catch 22. Conversely, carnivores are vulnerable to habitat fragmentation due to farming practices in rural communities because of their large home ranges, and sometimes their wide-ranging movements.

Predator control is one of the oldest, most widespread forms of wildlife management. The scale of some control programs is so extensive that it is easy to wonder how any large carnivores have survived! Predator control programs in the United States killed more than 286 000 large (≥9 kg) carnivores in 1998. This was three years after gray wolves were reintroduced to Yellowstone National Park, Wyoming, to *reverse* the effects of their extinction in the region. In Australia, approximately $9.5 million was spent on dingo control during 2007–2008. In NSW, $1.41 million alone was spent in 'conservation areas'. This is in contradiction to the management objectives of the NSW *National Parks and Wildlife Act 1974* that include:

- protection and preservation of scenic and natural features;
- conservation of wildlife and natural biodiversity; and
- maintenance of natural processes as far as is possible.

Once I was surprised to learn during a presentation by a Victorian scientist that showed the $200 000 worth of stock losses in one year was much less than the millions of dollars spent on dingo control. Results from a survey that investigated the economics of livestock attacks by 'wild dogs' in Queensland during 2009 suggested that wild dogs can cost up to $67 million in stock losses. But the survey participants were requested to estimate the monetary values that they could place on stock losses on site and at the sale yards. The resultant media frenzy, however, only reported the $67 million setback with no synthesis of the survey, other results or other issues. A compounding effect is that the market value of livestock varies on a daily basis, and there are no systems in place to quantify the effect of dingo predation.

Lambing rates and calving rates can be used as a baseline to investigate effects of predation if all livestock were pregnancy-tested. An overall annual loss of livestock may then be calculated at the end of the season when remaining livestock are herded. These rates are affected by the breed, genetics, seasonality, body weight, diet and nutrition, and even disease, vitamin deficiency or other environmental variables such as drought, fire and floods. Therefore, only if these factors are known when livestock productivity measures are determined for lambing rates and calving rates, and annual attrition rates of breeding stock, can dingoes be held potentially responsible for part or all $67 million worth of damages. In an interview for *Meat Trade News Daily* on 13 December 2009, Tim Schatz from the Northern Territory Department of Regional Development Primary Industry, Fisheries and Resources stated:

> 'When you've got 20 000 cows spread over thousands of square kilometres it's hard to know what's going on. Nobody actually knew what it (the return in calf rate) was until we actually went out there and measured.'

That statement throws even more doubt over the apparent $67 million setback. In another interview, this time on the ABC's *7:30 Report*, scientist Lee Allen from the Queensland Department of Primary Industries stated that there is *anecdotal evidence* suggesting one or two dogs can kill 50 or 60 sheep in one night. He also showed that one dog travelled 600 km from Charleville in Queensland, through the dingo fence to sheep property in NSW, much to the astonishment of the property owners! Does that mean that they hadn't lost 50 or 60 sheep in one night, even in the presence of dingoes? Or was that the first dingo to colonise their area since the dingo fence had been constructed? They obviously had not been implementing any dingo control owing to their being unaware of the presence of dingoes on their property. Here we have further evidence that dingoes don't migrate back and forth between livestock enterprises, mountainous country or flat country, despite the numerous claims that they do.

Kim Berger, an American scientist, questioned the efficacy of long-term efforts by the US Government to improve the viability of the sheep industry by reducing predation losses. Using data compiled from 1920–1998 she researched associations between changes in sheep numbers and factors such as predator control effort, market prices and production costs. Her results indicated either that predator control was ineffective at reducing predation losses or that factors other than predation accounted for the declines in both regions. The five factors that were most closely associated with decline in sheep numbers were:

1 average lamb price per 100 pounds received by ranchers;
2 average hay price per tonne;
3 average hourly wage rate paid to field/livestock workers;
4 percentage of sheep ranchers aged 65 and over; and
5 federal and cooperative dollars spent on livestock production.

Combinations of these top five factors were dominant in 14 of the 16 models used to estimate annual changes in sheep numbers. The next most dominant factor was the price of shorn wool. In addition there was a strong correlation showing that trends in sheep numbers continued to decline during the periods when 1080 bait was used to control coyotes and after 1080 use had been stopped due to concerns for animal welfare. In conclusion, Kim stated:

> 'That control efforts have had little effect on trends in the sheep industry is remarkable given the enduring nature of the program, the considerable resources devoted to carnivore removal (about $1.6 billion in real dollars between 1939 and 1998), the number of carnivores removed, and the frequent assertion that federal control of predators is necessary to maintain the sheep industry … If predation losses are the primary cause of the sheep industry's decline, then control, as practiced, has not been successful at reducing predation losses to the level necessary to make sheep ranching economically viable. Either the reduction in carnivore abundance has not been sufficient to produce effective results or the relationship between carnivore removal and predation losses is tenuous … Alternatively, control efforts may be highly effective at reducing predation losses, but the financial effect of losses may be secondary to other economic considerations. For instance … for a large percentage of producers, lamb and wool production is not profitable even in the absence of predation losses.'

The Australian sheep industry has apparently suffered continuously from restructuring. The core problem identified for this tendency was a re-weighting of the value derived from the sheep and lamb meat industries relative to the value derived from wool. Due to this, the national sheep flock which comprised 170 million head in 1990 declined dramatically to 76.9 million head in 2008 – the lowest it has been since 1916. That 55% rate of decline occurred at approximately 4.2% per year, or around 4 million head per year. Key changes identified by scientists in the industry included:

- number of sheep and producers in Australia has fallen and continues to fall;
- mix of sheep has changed with ewes dominating in a way not seen before;
- diameter of the wool being produced, which affects its market value, has fallen over the last 15 years; and

- gross value of product for sheep is equivalent to the gross value of wool.

Based on recent performance, projections have suggested that the Australian sheep population may well continue to decline. In addition, the number of lambs born in recent years has been less than the total of lambs slaughtered, sheep slaughtered and live sheep exports, and the population is effectively unstable and unsustainable.

Funnily enough, the notorious, infamous dingo in this setting has nothing to do with the market value of wool. Global production, supply and stocks of wool have also decreased since 1990, and the wool financial markets have been affected by variation in the global economy. These aspects provide a clear indication of how livestock industries are affected by more dynamic factors other than predation.

In the 2005 *Balancing Act* report by the Australian Federal Government, a 'generalised input-output analysis' was used to develop a numerate triple bottom line account of the Australian economy. The sheep and shorn wool sector actually has relatively large resource requirements. Greenhouse emissions, water use and land disturbance were all substantially above average in comparison with other sectors. The sector dominated with 94% of the greenhouse emissions, 78% of water use, and 100% of land disturbance indicating that the whole sector needs improvement from within. Of the sector's greenhouse gas emissions, fuel use produced 2% and 97% was due to methane from the digestion processes in sheep, with higher levels of emissions from sheep grazing poor quality pastures. This may be exacerbated by merino sheep because they dominate the Australian sheep flock (85%) and fine wool merinos need low quality food rations to produce fine wool. Other contributing factors include irrigation for pasture and crop that is used to fatten lambs and to accelerate live weight gain in ewes to stimulate ovulation and subsequent pregnancy. Feedlotting of sheep was also suggested to raise issues in consumer acceptance between 'lambs from grass' and 'lambs from lots'. Opposing this view are data that show livestock in feedlots produced less methane than livestock grown on grass. *Livestock's Long Shadow*, a report by the United Nations, raised similar concerns for the livestock sector at a global scale. In calculations of the impact that this sector has had on the environment, the scientists included issues related to land degradation such as deforestation, erosion and salinity; water pollution such as eutrophication; air pollution such as methane emissions; and loss of biodiversity.

Despite these issues, it is important to remember that many small town local economies are based on the livestock industry and agriculture in general. In some cases, especially in east and south-eastern Australia, livestock enterprises often run adjacent to dingo habitat and dingo conservation zones. These are the places that feel most threatened by dingo predation. Any state, national and global economic issue and most major environmental issues may appear worlds away from their enterprises but these issues are the nature of the business. Major research questions in the future should aim to identify how to adapt local enterprises to changing economic markets and to changing landscapes. Basing land management around ecosystem function and adapting livestock production to suit the local environments will supposedly maintain the viability of the local communities that the livestock industries are based upon, and the local landscapes.

The shark of the Australian bush

Anyone who has ever been surfing, snorkelling or spent time out in the ocean would have given thought as to what lurks beneath. It is the great unknown. Similar thoughts may eventuate when people are standing at a vista looking out over the Australian bush. Some days it appears to be so still and desolate. On one hike to retrieve a GPS collar in a remote location we were reminded pretty quickly that we weren't alone when we happened upon a swamp wallaby carcass that was less than a few days old. The dingoes had bailed it up against a small, eroded wall of a dry creek bed. Then when we arrived at a moist creek bed and saw dingo tracks in the mud, those feelings of being watched were reinvigorated. There is no doubt in my mind that dingoes knew exactly where we were, and probably even watched us walk past.

On another occasion with volunteer Stu Sutton, we had walked for a couple of hours to retrieve a VHF collar that had been omitting a mortality signal for a while. The area where *Binure* had died was a grassy flat between two creeks, next to quite a large wombat burrow. Stu was gathering *Binure's* bones and putting them in a pile together, when we heard a pup howling up a hill. So we ran up the hill towards the howl and heard a quick shuffle in the bush, followed by a bird call. We had been tricked by a lyrebird. The fact that the lyrebird was impersonating pups makes me suspect that the wombat burrow was actually a den site.

On some hikes in the most remote places I still find signs of dingoes – tracks, scats or prey carcasses. Sightings of dingoes, however, are uncommon. In some instances a dingo spotted on a fire trail quickly vanishes into the bush because they can stand motionless and remain undetected. The tan and black colours blend perfectly in with the surroundings – the same as every other predator. Sharks, lions, tigers and wolves are all hard to see because if they were discernable from other features in their habitat, it would be too easy for prey to see them and escape.

If the dingo wants to be seen it will make itself known. The most memorable occasion for me occurred while checking a sand plot along the Wollondilly River. I howled to my sister who had stayed at the car and a moment later I heard some whimpering from behind a nearby bush and a dingo walked into view. To avoid intimidating the dingo I dropped to my knees and the dingo whimpered back and forth cautiously before it slowly slinked away. The scrub was too dense to follow it so I went back to the car and the dingo had instead followed me. He walked around the car and to an old farm dam where he started to howl.

On a field trip with freelance reporter Erin O'Dwyer, we saw *Bunyal* (sun) have a drink at the top dam during an afternoon stakeout. She was emaciated and gave the impression that she was alone and nervous. On twilight an hour or so later, Erin requested we go for a short walk in hope of seeing more dingoes. After walking along a ridge and back down onto the road, it began to get dark so we decided it was time to walk back to the car. After walking back along the road we started to get near the car and Erin asked me to howl. Within 10 seconds, at least four, maybe even six dingoes howled back from the part of the road we had just walked. It was spine tingling but enlightening to know that these stereotyped 'deadliest' predators from Australia simply watched us walk past them.

Some people may have been very disconcerted being that close to wild dingoes. Scientists and naturalists on Fraser Island have been within a couple of metres of dingoes fighting and haven't been attacked either. Why are some people attacked and others not? If we assess previous dingo attacks then there are many similarities – mainly, the dingoes are habituated to the presence of humans and in some instances, they have become dependent on humans. In a large proportion of cases, the victims of attacks are children, or scared or nervous teenagers and adults. Obviously they weren't carrying their firestick! In areas like the Blue Mountains or the desert where interactions between dingoes and people are uncommon, the dingoes tend to run at the sight or scent of people. Dingoes that have frequent interactions with people on Fraser Island in Queensland or even at Seal Rocks in NSW behave differently around humans compared with other dingoes.

In addition, there are hundreds more attacks on people by domestic dogs annually than by dingoes. Statistics of reported dog attacks in NSW during 2004–2005 showed that one of 376 dingoes on the register had attacked a person. In comparison with other dog breeds that attacked people there were:

- 63 of 35 711 German shepherds;
- 59 of 28 850 Australian cattle dogs;
- 58 of 23 735 Rottweilers; and
- 33 of 3244 pitbull terriers.

Table 8.1 shows data on reported attacks by domestic dogs in NSW from 1996 to 2003. In data from 2004 to 2005, urban councils reported a total of 829 attacks while rural councils reported 44 attacks.

Unfortunately there is no synthesis of these data to show why people were attacked. Were the dogs neglected or trained guard dogs? Were the dogs defending themselves or being provoked? It is clearly evident, however, that the 2487 domestic dog attacks have received much less attention than the two or three dingo attacks across Australia over the same time period.

In a study in Adelaide, about 6500 people were injured annually by dog attacks. Children aged 0–4 were attacked and required hospital twice as often as adults aged 21–59 years. Head and facial bites were more prevalent (90%) in children aged less than 12 years old. Once again, dingoes were not mentioned as high risk, but German shepherds, bull terriers, heelers, Dobermans and Rottweilers were identified as the main offenders.

Why are dingoes treated so differently?

Table 8.1 Attacks by domestic dogs in NSW 1996–2003.

Years	Reported attacks	Attacks in public	Attacks on private property	Attacks on people	Attacks on animals
1996–2003	2487	1625	766	1273	840

9

CONSERVING DINGOES IN AUSTRALIAN LANDSCAPES

The IUCN Canid Specialist Group provides comprehensive reviews of canids to aid conservation through their Canid Action Plans. In the 1990 plan they stated that conflict from humans towards canids is based on the underlying assumption, rather than proof, of competition. If livestock were found in the diet of the canid in question, then the livestock could have just as easily been eaten as carrion or been sick. In both instances, the livestock would have already been valueless, like the measure of competition itself. Therefore, it is important to identify whether the predators are taking the doomed surplus or more. Three categories of information were recognised to assist in evaluating any damage by canids: 1) a measure of competition; 2) a measure of the loss; and 3) the consequences of action. These measures themselves may be subjective based on the objectives of the organisation or individual and how they value the stock.

Conflicting views about dingoes makes dingo control, preservation, conservation and management difficult. Dingo management therefore is a matter of perception and mainly influenced by the perspective of the beholder. If one thinks that pure dingoes exist and need to be saved from extinction then they also believe that all other forms of wild dogs or hybrids should be destroyed. Some government organisations are governed by legislation that states dingoes must be destroyed while others are governed by legislation that states dingoes are protected. Then there are issues with the control of dingoes because this has disturbed and continues to disturb their pack structure. Preserving dingoes alternatively removed them from their environment and caged them in zoos or dingo sanctuaries where they perform mainly an educational role.

What then is dingo management? It seems to be a collective agreement among land owners and land and wildlife management organisations to reduce the effects of dingoes on people. These may be exaggerated or they may be an excuse to account for other dilemmas associated with the economics of livestock production. Management of hypercarnivorous predators around the world is, however, based on the same attitude: 'if it affects me or my objectives then it should be killed.'

How true – survival of the fittest. But what contributes to fitness more than a healthy environment? Canids are landscape specialists and surely will adapt to some changes and disturbances in the environment better than humans. After all, they have survived in various forms for 10–12 million years and lived in the Arctic, the tropics and in arid lands. If arable land decreases because it has been cleared, grazed, degraded and modified to the point it is no longer arable, where will humans produce food for livestock and other humans?

In 1996, 2001 and 2006, national Australian State of the Environment reports indicated that land clearing and agriculture has had a more detrimental impact in 200 years than dingoes have had in 5000 years. Dingoes were only mentioned in the 1996 report in context of:

- the threat of hybridisation with domestic dogs and loss of genetic purity;
- the effects of dingo control in modified landscapes, which has exacerbated the negative effects of increased grazing pressure; and
- as beneficiaries of increased prey resources.

Agricultural and pastoral industries, however, were implicated in all three reports as environmental pressures with possible disastrous effects on Australian landscapes. Effects include extinction of 24 native mammal species, increased risk of dryland salinity and desertification. Short-term economic gain by pastoralists and government from dingo control also fails to account for the broader questions regarding carbon emissions or climate change.

Is perpetuation of dingo control for over 100 years solving the problem?

The answer to this is clear and simple - No. If dingoes are the problem then dingo control appears to exacerbate livestock predation. If livestock are the problem then dingo control inadvertently exacerbates land degradation. The impacts of dingoes on the livestock industry are minor in comparison with the impacts the livestock industry and agriculture in general has had on Australian landscapes. The considerable resources devoted to dingo control, the number of dingoes controlled, and the assertion that controlling predators is necessary to maintain the viability of the Australian livestock industry appears to be a façade.

My proposed hypothesis to counter the effects of dingo predation on livestock enterprises and the effects of livestock enterprises on Australian ecosystems is:

> If livestock enterprises monitor local dingo populations regularly, then they will be able to adapt management of their livestock to minimise predation by dingoes and reduce the impact of their livestock on Australian landscapes.

Australia includes many fragile ecosystems for which drastic measures are needed to reduce the impact of agriculture and livestock. Instead of funding dingo control, relevant industries could fund research to investigate ways to reduce their impact on Australian landscapes and ways of adapting their industry to suit Australian environments.

Management of dingoes should instead consist of adaptations to extant livestock husbandry practices, as opposed to dingo control. Humans are better at directing, administering and supervising other humans than they are at controlling dingoes and this is evident because there is a continuous lobby to control dingoes. Livestock enterprises affected by dingo predation should instead be encouraged to learn more about their local dingo population, in some cases as far as consistent monthly or seasonal monitoring of dingo activity on or adjacent to their properties. To assist entrepreneurs to manage their livestock accordingly, an audit of affected enterprises may be beneficial and recognise other factors such as:

1   where the enterprise is positioned in relation to dingo territories;
2   where and when the local dingoes breed;
3   how the activity patterns of dingoes fluctuate per season, including where activity is highest and lowest, on or adjacent to their property;

4   whether there are many attractants for dingoes such as unspayed
    domestic dogs or littering of livestock corpses and offal;
5   whether there are many deterrents for dingoes such as guard animals,
    shepherds or electric fences; and
6   whether optimal living conditions for dispersing/lone dingoes to
    inhabit are made available adjacent to the enterprise by control
    programs.

If dingo populations continue to be a problem after the enterprise has
complied with an audit for dingo attractants and deterrents, then targeted
culling by the appropriate authorities of troublesome dingoes involved in
persistent predation of livestock could be used as a final option. It is
recommended that this option be used only if all other techniques have been
exhausted.

Adaptive management

The current management system for dingoes around Australia is expensive,
time-consuming, possibly implemented for the wrong reasons and rarely
accounts for ecosystem function. The proposal below for improved
management intends to promote coexistence between farming enterprises
and the preservation of natural phenomena.

It is now understood that hypercarnivorous canids require communal
living to learn behaviours. Chapter 3 showed that dingo packs in the Blue
Mountains consisted of closely related genetic social units. Chapter 6
showed social groups are restricted to core areas of activity for 10 months
of the year. Seasonal variations in spatial organisation were consistent with
the seasonal variations in prey preference and patterns of activity, showing
the function of dingoes in Australian landscapes. The most plausible
explanation for dingoes to maintain this functional role is the maintenance
of pack structure and social systems.

Destruction of native fauna due to community pressure does not appear
to be in the best interest to achieve the goals set by most land management
organisations. The best-practice method of management then is to cease
attempting to 'control' dingo populations because they may be self-
regulatory. Instead, management should review farming practices on
properties affected by dingoes to minimise interactions between dingoes and
livestock. Circumstances to be reviewed are those which may attract a dingo

to a property and those which may deter a dingo from a property. If there are more attractants than deterrents then it is not a land management problem, but rather a farm management problem.

This form of management can work in two stages. From June until November, encompassing low periods of dingo activity (whelping and rearing seasons), properties can be audited for dingo attractants and deterrents. These can be resolved in time for the training and breeding season when range expansion of dingoes is more extensively documented. The following six months (December–May) of higher dingo activity and stronger social ties can then be monitored for comparison with the preceding six months. During this period, however, livestock producers could alter methods of managing their livestock by learning more about the biological seasons of dingoes in their area and adopting practices and breeding regimes to reduce the risk of attack to livestock. Moving the stock closer to the residence or into a dingo-proof section of the enterprise would be recommended at prescribed times of the year. Public property, including school yards and camping grounds, on the contrary, will require more regular removal of dingo attractants, especially food and litter.

Land management organisations would need to provide collaborating pastoral enterprises with quality assurance that they meet the requirements of the audit. Any subsequent loss of livestock should result in a consultation to ascertain reasons for these losses, amending farm management practices if necessary and targeted management of the rogue dingo or dingoes that have learnt to attack livestock. Using this method may reduce expenditure on widespread dingo control campaigns, even if livestock producers were compensated or subsidised for their livestock lost to dingo attacks. In a comprehensive review of predator management techniques, scientists suggested the majority of lost livestock in Australia were not due to predation but other biological factors, which also need to be accounted for during such audits, such as the price of wool and meat, and the effects of drought and flood.

Collaboration between scientists, farmers and land managers is integral to the success of the management technique outlined, and has the ability to influence change. This form of management has many levels and topics that cannot be entirely addressed in this book. Land management organisations could employ a farm sustainability officer to work not for or against but *with* each enterprise. The most important concept in contention is that many tiers of culture will have to accommodate change and evolve to a new level.

Fortunately this cultural evolution can be based on objective scientific measurement rather than on subjective judgement, anecdote and the resultant cultural transmission of behaviours between humans. Government organisations will need to review and alter policy to suit the needs of the environment in harmony with the needs of the populace. In the context of sustainability and ecologically sustainable development, the long-term benefits will outweigh the short-term gains that contemporary dingo management is directed towards.

In the current structure of organisations charged with the responsibility to manage Australian landscapes, there appears to be division between the decision-making process and the implementation of management decisions. Local management can sometimes be diverted to servicing the needs of local communities while attempting to maintain the values outlined by decision makers for the global interest groups. The goals of interest groups may conflict and adaptive local management practices may outweigh long-term strategic planning. On one front, the globally iconic dingo is an essential component of Australian landscapes. On the local level, dingoes are often seen as vermin that require control.

Non-lethal management techniques

There are a number of non-lethal management techniques either available or being developed that effectively limit the possibility for conflict in the first place. Improvements in animal husbandry may not be costly and may return significant results. Other techniques include:

- proper disposal of carcasses;
- confining animals at night, when they have young/infants or when predators are more active;
- shepherding;
- guard animals (see Figure 9.1);
- predator fertility control;
- repellent chemicals;
- compensation and cost-sharing schemes;
- electrical or predator-proof fencing;
- conditioned taste aversion;
- shock collars;
- aversion and disruptive stimuli;

- improved animal husbandry;
- improved reserve design; and
- bio-boundaries.

Some of these techniques such as using guard animals like alpacas and maremma dogs are slowly infiltrating management practices in the Australian livestock industry. Many options, however, are infrequently explored. Shepherding is one technique that could be reintroduced at certain times of the year. This may stimulate local and state economies by attracting workers, redistributing funds to individuals and the millions of dollars currently spent on control programs could instead be spent on other forms of employment. A shepherd may also double as a sustainability officer and perform the audits for dingo attractants and deterrents, and provide advice for adaptive livestock husbandry and adaptive management.

Figure 9.1 A guard dog guarding a flock of goats from my parents while they were trekking in Spain. Apparently another guard dog was with another flock of goats in the valley below, and yet another guard dog had sniffed them out and was watching from behind. They decided to take one photo and keep walking. Image: Bryan Purcell

Bio-boundaries are a new technique being developed in Africa to limit the movements of African wild dogs. Chemical markers found in the urine and anal gland secretions of individuals play a role in canid social systems and have been used to mock a territory. Faeces of dominant wild dogs were placed in strategic locations at the boundaries of reserves and livestock enterprises and some travelling wild dogs were deterred from the area by the scent.

As alluded to earlier, it is highly important for dingoes and other canids to maintain scent posts. If it is the scent of dominant animals that deters travellers from entering certain areas or if it is the combination of scent by multiple pack members, use of either technique could form a new, non-lethal management option. It may be that the frequency of scent posts increases to the point that it is beyond doubt an intruder has entered or is entering the core area of a pack. Livestock producers may be able to imitate a dingo home range by increasing the frequency of scent posts and patrolling borders regularly with their domestic dogs.

Another option open to farmers could be the construction of buffer zones on their property. Parts of the property adjacent to dingo habitat could have less livestock than areas further away from dingo habitat. If they farm sheep and cattle, then they could put the cattle in yards adjacent to dingo habitat and then the sheep in yards further from dingo habitat. At my study site, pine plantations appeared to provide a buffer between livestock enterprises and preferred dingo habitat in some areas. Adaptive management is only really limited by imagination.

Case study: the Murray–Darling Basin

The Murray–Darling Basin (MDB) is the only catchment that is being governed at a federal level and the only area in Australia where dingoes have been excluded. About 85% of the MDB is inside the dingo fence. Approximately 67% is used for agricultural purposes and this proportion comprises Australia's 'food bowl'. Land use in the area is predominantly irrigated crops including pasture (43%); cereals other than rice (20%); cotton (15%); rice (6%); grapes (6%); fruit and nuts (5%); and vegetables (2%). A comprehensive report by the Australian Bureau of Statistics identified the following as issues of priority in the 19 Natural Resource Management (NRM) regions:

- water quality and/or quantity (identified by 16 of the 19 NRM regions);
- salinity (irrigation and dryland) (identified by 14 of the 19 NRM regions);
- biodiversity (identified by 14 of the 19 NRM regions);
- soil health and/or soil erosion (identified by 10 of the 19 NRM regions);
- native vegetation (identified by 9 of the 19 NRM regions); and
- weeds and/or pests (identified by 8 of the 19 NRM regions).

This is a very diverse region with dramatic variations in climate, topography and temperature, and many environmental problems that will have worse effects on agricultural productivity than dingo predation. A disproportionate amount of funds was spent on weeds and pests than on other issues. Some scientists have labelled the area a conservation wasteland and have suggested reintroducing dingoes to limit the abundance of foxes and cats and control grazing pressure by overabundant herbivores. It is hoped that the dingoes will restore some natural balance to the landscape and help it to regenerate.

Research on the effects of wolf reintroduction to Yellowstone National Park showed that overpopulating ecosystems with herbivores not only has detrimental impacts on the ecosystem, but also that detrimental impacts can be naturally controlled. The MDB appears to be in a worse state than Yellowstone National Park. Dingoes have been excluded from this Basin since 1885 by the longest fence in the world for livestock farmers to maximise economic benefits from their livestock. The impacts of agriculture are well-documented in the MDB and establishment of a federal commission to manage the unstable ecosystem is further evidence of its poor condition. Although the removal of the dingo fence to rejuvenate the MDB has been suggested, in its present state with low water flows and saline soils, a reduction in herbivore abundance would be ineffective without landscape remediation work by humans. Rehabilitation of the natural ecosystem and reduction of unsustainable agricultural practices should be a focus for adaptive management of the MDB and other Australian landscapes. Such a strategy may lead to reintroduction of dingoes to maintain biological stability and repair the integrity of some Australian ecosystems.

If dingoes were to be reintroduced, then livestock producers and land management organisations will first need to collaborate for an effective adaptive management strategy. Dingo reintroduction would need to be implemented slowly, where one pair are released in an area with less livestock enterprises. Livestock enterprises in the meantime can begin to prepare their property for coexistence with dingoes such as establishing mock dingo territories. Following the slow release of dingoes into the area, the formation of packs and territories, the fence may slowly be opened and eventually removed.

Predation of livestock may still occur and may be worse in some places than others. Monitoring and managing outstanding issues at the local level, however, will prove more effective than managing issues across regions or across the nation. There were variations in diet and activity of dingoes between two sites that were only 20 km apart in the Blue Mountains. In addition, some property owners had more problems with livestock predation than others, but this may have been because more control programs were implemented in their area. This was consistent with research showing that predator control has had little effect on the viability of livestock industries in America and in Australia. The most important issue is that the Australian environment is cared for first, because sustained environmental degradation will only limit agricultural productivity and our enjoyment of the Great Southern Land.

10

ORDER IN THE PACK

Stories of the dingo generally take two forms, which largely depend on the story teller. One form is a romanticised narrative from people impressed by the wild heart of the Australian bush. The other form of story is a yarn by a classic 'Aussie battler' at war with a cunning, enigmatic enemy and usually ends in bloodshed and a trophy pelt or scalp. Dingo stories have been preserved by writings and artwork and explain what it may be like to be a dingo, or to be living in Australia with dingoes. One common theme in dingo stories is that the iconic Australian is a scapegoat. Part of the public reaction to the cry of Lindy Chamberlain on the 17 August 1980 at an Uluru camping ground, *'the dingo's got my baby!'* may have been because they felt the dingo was being made a scapegoat again. Regardless of how the baby Azaria Chamberlain died, Lindy Chamberlain was incarcerated by the resultant media and public outcry for being an uncaring parent. Similarly, public outcry could inhibit pastoralist activities for saying something such as *'dingoes killed my sheep!'*

Dingoes are an esteemed Australian mammal yet the largest predator-proof fence in the world was constructed to limit their distribution for the benefit of the agricultural sector, without exploration of the impact that exclusion of dingoes might have on semi-arid Australian ecosystems. Early Australian development relied heavily on the sheep industry and protection of sheep was deemed to be more important than the protection of dingoes. Declines in biodiversity including the extinction of 24 native mammal species that represented over half of the global mammal extinctions in 200 years could be attributed to the rise of agriculture and the neglect of ecosystem function. High-density dingo populations may aid the conservation of native Australian marsupials. It seems strange then that public outcry against the suppression of dingoes has not questioned the validity of the dingo fence and dingo control programs. Researchers have shown, however, that livestock farmers are protected by such public outcry due to Australian frontier nationalism, a culturally inherited trait of many Australians that protects 'Aussie battlers' using the prejudices inherited from European colonisation.

While studying the dingoes in the Blue Mountains, a male/female dingo pair and two of their pups had been attracted to a local school. The father and pups were inevitably controlled by staff from land management organisations because they were a threat to the safety of the students; but the dingoes were cleaning the rubbish out of the yard! Response to an actual fatal incident on Fraser Island during 2001 was lethal management of some dingoes followed by trials of non-lethal techniques under the Dingo Management Strategy formulated in response to the event. Prior to the implementation of a control program, the first response to any statement blaming a dingo or dingoes for prevailing circumstances or events should be *'why?'* For instance, ascertaining reasons why or how a dingo would be prompted to seize a baby from within a tent in a camping ground, could be related to cultural transmission of behaviours in an area where dingoes have been habituated to obtaining food in camping grounds. This is similar to dingoes foraging in school grounds or livestock enterprises located adjacent to dingo habitat. The dingo is innately a predator first and this should be considered in development of management strategies for landscapes where dingoes and humans intersect.

Canids learn behaviours through cultural transmission. Cultural transmission occurs when behavioural traits are inherited (vertical transmission) from parent populations or copied (horizontal transmission) from neighbouring populations. Scientists have demonstrated how domestic dogs learned behaviours by watching their mothers, conspecifics or a human equivalent. Problem-solving behaviours in domesticated dogs involved complex social learning interactions and individual experiences. Asocial (individual) and social learning abilities apparently contribute to increased fitness of individuals. Observational learning thus becomes the nexus for cultural transmission of behavioural traits such as predation on livestock, predation on native species or scavenging from human settlements.

The above observations imply that the dingo at Uluru learnt, through observation, that exploring inside a tent would result in the opportunity of taking food. Dingoes have been known to become habituated to finding food items around camp sites and this may have been reinforced by campers willingly offering food items to dingoes at the time of the event. In contrast, to domestic dogs, wolves like dingoes are naturally skilled problem solvers, highly social predators with a well-organised social hierarchy who depend on vertical transmission to acquire behavioural traits.

Australian scientists have explored reasons why livestock losses increased following dingo-baiting programs. They suggested that the effect was caused by lost social ties in dingo groups and changes in the age structure of the dingo population, but how does a juvenile dingo know livestock are prey? Livestock predation may result from the collective influence of asocial learning, opportunities, and motivation (hunger) but horizontal transmission cannot be overlooked. Imitation behaviours are apparently distinct from asocial and social learning and dingoes may watch working dogs herding livestock. Comparable with natural selection and behavioural adaptation, imitation behaviour may incite cultural evolution that inherently causes imitation behaviours.

Domestic working dogs are part of the moral code and right of passage for livestock farmers in Australia. They serve many functions, one of which is to herd livestock. Although herding behaviours may be instinctive for dingoes, observational learning may be a more instinctive behaviour than the imitation or action of behaviours in domestic dogs. This herding task could be observed and learnt by dingoes, especially juvenile loners with deceased parents, or others with limited social ties. Considering dingoes are predators they may have innate herding instincts and their cryptic habits are an indication that working dogs may not detect their presence easily. Livestock farmers have previously described through poetry how dingoes drove sheep, which is consistent with imitation of sheep-herding behaviours. Current dingo management practices of creating a dispersal sink or buffer zone adjacent to livestock enterprises is the most likely cause for observational learning by roaming dingoes because optimal habitat is made available for habitation. Using satellite/GPS telemetry, researchers in Queensland have showed that a lone dispersing dingo rapidly inhabited a recently baited buffer zone. Imitation of domestic working dog herding behaviours by dingoes toward livestock may be the reason behind some predation on livestock. Some dingoes may even be attracted to livestock properties because of a female domestic dog in oestrus during the breeding season.

The confrontation between humans and dingoes exists because of the flaws of humans. Australian frontier nationalism is one of those flaws – a culturally inherited patriotic trait of many Australians. Some perceptions were that the Australian bush was a place for men, being too scary and too dangerous for women. This culture has been suggested as part of the reason Lindy Chamberlain was imprisoned and it is this same culture that imprisons dingoes. The dingo could have taken Azaria, but intentionally or not, Azaria was taken to the dingo. In the same sense, dingoes may predate livestock, but intentionally or not, livestock are taken into areas that are traditional dingo habitat. These and other causes of conflict between dingoes and humans are the responsibility of humans and not dingoes. This understanding is vital in developing a novel approach for dingo management.

Who's afraid of the big bad wolf?

Recent research has attempted to identify the social impacts of dingoes on livestock producers. Using the Upper Hunter Valley in NSW as a case study, the report showed that both the number of sheep and cattle producers had declined in the region as a result of changes in the industry. Wild dogs or dingoes were also implicated for reduced farm income, financial stress and additional expenditure, loss of community cohesion, land-use change and psychological distress. Apparently wild dogs, 'especially dingo crossbreeds', also undermined the sustainability of sheep farming.

These results are in contradiction to data presented in Chapter 8 that showed global financial markets had a bigger impact on sheep farming than predation. Psychological distress of seeing mauled livestock will obviously play a role in attitudes towards dingoes. If the livestock producers had a better understanding of the local dingo population, then they could adapt their livestock husbandry and reduce the risk of dingo predation. Attempting to implement community-wide pest control programs was identified as a social issue for communities, because some landholders have different values to livestock producers. All sentiments therefore imply that improvements in livestock husbandry may be more beneficial for individuals and communities.

Some scientists also have raised alternative arguments regarding dingo reintroduction in the Murray–Darling Basin, due to the changing nature of wild dog populations. This book has presented evidence showing that wild dog populations may not actually be changing in nature but landscape management practices should. Numerous studies have shown that dingoes and/or wild dogs, in comparison with foxes, had a negligible effect on populations of native Australian endangered species. The only apparent reason to control dingoes, therefore, appears to be for the protection of livestock enterprises. Chapter 8, however, showed that government-subsidised predator control failed to prevent the decline in the sheep industry in the USA. Evidence presented instead suggested other economic variables influenced the viability of livestock industries more than predation. Predators were, however, held responsible and essentially were blamed for the faults of humans.

A new understanding of the dingo

In the end, the dingo is simply trying to survive by maintaining acquired territory or by acquiring new territory. Humans have the capability and the doggedness to adapt to different landscapes so easily that it is probably time to live within the Australian biome, rather than against it. Minimising conflicts of interest, whether they be cultural or financial, will open a new era for the science behind dingo management. It is anticipated that adaptive management of livestock to dingo biology will reduce conflict with dingoes, assist objectives for dingo conservation and improve biological stability in degraded Australian ecosystems. With appropriate management regimes in place, then the potential exists to reintroduce the top-order mammalian predator for maintenance of biodiversity in Australian landscapes from where they are currently excluded (see Figure 10.1).

Parallels of natural resource management exist also between Australia and America. Recent Australian conservation values for dingoes appear to have changed as a consequence of the American conservation values for wolves. Wolf biologist David Mech has identified numerous factors about the value of wolves in American ecosystems concluding that wolves are one of the top-down regulating mechanisms that maintain biological stability. Since their reintroduction into Yellowstone National Park, the wolf

Figure 10.1 Are these the sign of a stable ecosystem? Typical track marks left by a dingo that crossed a sand plot. Image: Aran Anderson

population has grown to the point that wolf management needs to be re-evaluated due to increasing attacks on livestock on adjacent ranches. More to the point, a level of ecosystem stability was regained from the reintroduction of the higher order predator.

Dingo management in NSW seems to be influenced by the wolf reintroduction strategy. In 2000, an internally generated submission for a second wild dog control order under the *Rural Lands Protection Act 1998* was prepared by land management organisations to declare some public lands in NSW important for the survival of dingoes. This move marked the beginning of a cultural revolution recognising the importance of the dingo to Australians and the biological stability of Australian ecosystems. Many current scientific studies are investigating diet, movements, improved management, people's perceptions and the behaviour of dingoes. There is so much interest in this topic that our understanding of dingoes is constantly increasing. For instance, on a field trip in September 2009 while I was writing this book, four dingoes chased a kangaroo into water and continued to attack the kangaroo while they were swimming. Until that moment, all scientists believed that kangaroos used water to escape predation. Now it seems as though some dingoes have learnt to use water as a tool when hunting kangaroos.

A cultural revolution for Australians

The process of cultural evolution in Australia has commenced. Climate change is seen as an impending doom and many Australians are seeking reasons why it is a threat and how we can adapt. The significance of the impacts of livestock production systems on global climate is now well documented. Cultural traits, like those associated with climate change, are subject to intra- and intergenerational variations. Traits may conflict with one another but favourable heritable variants accumulate and culturally dominate over time. The land management organisations and landholders can show by example that dingoes and contemporary Australian people can coexist in a sustainable manner. It is highly recommended that adaptive management to promote coexistence with dingoes be publicly trialled. The example of dingoes may achieve other cultural shifts in thinking for future management of Australian landscapes and ecosystems. Perhaps wild canids such as dingoes and wolves will attain the same status as sharks internationally, so that they can continue to perform their key functional role as top-order predators in unregulated ecosystems. Such a change in culture may be essential if pastoral industries and natural ecological systems are to become sustainable in Australia.

REFERENCES

Abrams PA (1992) Adaptive foraging by predators as a cause of predator-prey cycles. *Evolutionary Ecology* **6**, 56–72.

Adams CE, Hamilton DJ, Wilson IMAJ, Grant A, Alexander G, Waldron S, Snorasson SS, Ferguson MM and Skúlason S (2006) Does breeding site fidelity drive phenotypic and genetic sub-structuring of a population of Arctic charr? *Evolutionary Ecology Research* **20**, 11–26.

Allen BL (2006) The spatial ecology and zoonoses of urban dingoes – a preliminary investigation. Honours Thesis, University of Queensland.

Allen BL and Miller HA (2009) *The biodiversity benefits and the production costs of dingoes in the arid zone: summary of research results for 2008.* South Australian Arid Lands Natural Resources Management Board, Port Augusta.

Allen JJ, Bekoff M and Crabtree RL (1999) An observational study of coyote (*Canis latrans*) scent-marking and territoriality in Yellowstone National Park. *Ethology* **105**, 289–302.

Allen L (2000) Measuring predator control effectiveness: reducing numbers may not reduce predator impact. In: *Proceedings of the 19th Vertebrate Pest Conference.* (Eds TP Salmon and AC Crabb) University of California Davis, San Diego.

Allen L and Sparkes EC (2001) The effect of dingo control on sheep and beef cattle in Queensland. *Journal of Applied Ecology* **38**, 76–87.

Allen L, Engeman RM and Kruper H (1996) Evaluation of three relative abundance indices for assessing dingo populations. *Wildlife Research* **23**, 197–206.

Allen LR and Gonzalez A (1998) Bating reduces dingo numbers, changes age structures yet often increases calf losses. *11th Australian Vertebrate Pest Conference.* Bunbury, Western Australia.

Anonymous (1978) Australian vertebrate pest control conference 1978: working papers. *Australian Vertebrate Pest Control Conference Proceedings*, Canberra.

Anonymous (1996) *Australia: State of the Environment 1996.* Independent Report to the Commonwealth Minister for the Environment, CSIRO Publishing, Canberra.

Anonymous (2000) *Submission for proposed amendments to the* Rural Lands Protection Act 1998 *Wild Dog Control.* NSW National Parks and Wildlife Service, State Forests of NSW, Department of Land and Water Conservation and Sydney Catchment Authority, Sydney.

Anonymous (2001a) *Blue Mountains National Park Plan of Management.* NSW National Parks and Wildlife Service, Hurstville.

Anonymous (2001b) *Kanangra-Boyd National Park Plan of Management.* NSW National Parks and Wildlife Service, Hurstville.

Anonymous (2001c) *Nattai Reserves: Nattai National Park, Bargo State Recreation Area, Burragorang State Recreation Area, Nattai State Recreation Area, and Yerranderie State Recreation Area Plan of Management.* NSW National Parks and Wildlife Service, Hurstville.

Anonymous (2001d) *Australia State of the Environment 2001.* Independent Report to the Commonwealth Minister for the Environment and Heritage, CSIRO Publishing on behalf of the Department of the Environment and Heritage, Canberra.

Anonymous (2006) *GPS tracking systems: state of the art GPS collars for wildlife tracking, wildlife monitoring systems.* http://www.environmental-studies.de/Argos/A-0.html [Accessed: 10.08.2006].

Anonymous (2007) *Australian Dingo.* http://www.ankc.org.au/home/breeds_details_print.asp?bid=103 [Accessed: 12.12.2007].

Anonymous (2007) *Council reports of dog attacks in NSW July 2004 – June 2005.* NSW Department of Local Government 2007, Nowra

Anonymous (2008) *2008 IUCN Red List of Threatened Species,* Canis lupus *ssp.* Dingo. www.iucnredlist.org [Accessed: 28.10.2008].

Asa CS, Mech LD and Seal US (1985) The use of urine, faeces, and anal-gland secretions in scent-marking by a captive wolf (*Canis lupus*) pack. *Animal Behaviour* **33**, 1034–1036.

Ashe A and Whitelaw E (2006) Another role for RNA: a messenger across generations. *Trends in Genetics* **23**, 8–10.

Atchley WR, Logsdon T, Cowley DE and Eisen EJ (1991) Uterine effects, epigenetics, and postnatal skeletal development in the mouse *Evolution* **45**, 891–909.

Atkinson RPD, Rhodes CJ, Macdonald DW and Anderson RM (2002) Scale-free dynamics in the movement patterns of jackals. *Oikos* **98**, 134–140.

Baenninger R (1978) Some aspects of predatory behaviour. *Aggressive Behaviour* **4**, 287–311.

Baker PJ, Boitani L, Harris S, Saunders G and White PCL (2008) Terrestrial carnivores and human food production: impact and management. *Mammal Review* **38**, 123–166.

Balshine-Earn S, Neat FC, Reid H and Taborsky M (1998) Paying to stay or paying to breed? Field evidence for direct benefits of helping behaviour in a cooperatively breeding fish. *Behavioral Ecology* **9**, 432–458.

Bandeira de Melo LF, Lima Sábato MA, Vaz Magni EM, Young RJ and Coelho CM (2007) Secret lives of maned wolves (*Chrysocyon brachyurus* Illiger 1815): as revealed by GPS tracking collars. *Journal of Zoology* **271**, 27–36.

Banks PB, Newsome AE and Dickman CR (2000) Predation by red foxes limits recruitment in populations of eastern grey kangaroos. *Austral Ecology* **25**, 283–291.

Baverstock P and Green B (1975) Water recycling in lactation. *Science* **187**, 657–658.

Beeton RJS, Buckley KI, Jones GJ, Morgan D, Reichelt RE and Trewin D (2006) *Australia State of the Environment 2006.* Independent report to the Australian Government Minister for the Environment and Heritage, Department of the Environment and Heritage, Canberra.

Bekoff M (1987) Group living, natal philopatry and Lindström's lottery: it's all in the family. *Trends in Ecology & Evolution* **2**, 115–116.

Bekoff M and Wells MC (1981) Behavioural budgeting by wild coyotes: the influence of food resources and social organization. *Animal Behaviour* **29**, 794–801.

Berger K (2006) Carnivore-livestock conflicts: effects of subsidized predator control and economic correlates on the sheep industry. *Conservation Biology* **20**, 751–761.

Blaum N, Engeman RM, Wasiolka B and Rossmanith E (2008) Indexing small mammalian carnivores in the southern Kalahari, South Africa. *Wildlife Research* **35**, 72–79.

Blewitt ME, Chong S and Whitelaw E (2004) How the mouse got its spots. *Trends in Genetics* **20**, 550–554.

Boitanni L, Francisci F, Ciucci P and Andreoli G (1995) Population biology and ecology of feral dogs in central Italy. In: *The Domestic Dog: Its Evolution, Behaviour, and Interactions with People.* (Eds J Serpell and P Barrett) Cambridge University Press, Cambridge.

Borgerhoff Mulder M, Nunn CL and Towner MC (2006) Cultural macroevolution and the transmission of traits. *Evolutionary Anthropology* **15**, 52–64.

Bowen WD and Cowen IM (1980) Scent marking in coyotes. *Canadian Journal of Zoology* **58**, 473–480.

Breckwoldt R (1988) *The Dingo: A Very Elegent Animal.* Angus and Robertson, North Ryde, NSW.

Brunner H, Amor RL and Stevens PL (1976) The use of predator scat analysis in a survey at Dartmouth in north-eastern Victoria. *Australian Wildlife Research* **3**, 85–90.

Brunner H, Triggs B and Ecobyte Pty Ltd (2002) *Hair ID: An Interactive Tool for Identifying Australian Mammalian Hair. CD-ROM,* CSIRO Publishing, Collingwood.

Burbidge A and McKenzie NL (1989) Patterns in the modern decline of Western Australia's fauna. *Biological Conservation* **50**, 143–198.

Burrows ND, Algar D, Robinson AD, Sinagra J, Ward B and Liddelow G (2003) Controlling introduced predators in the Gibson Desert of Western Australia. *Journal of Arid Environments* **55**, 691–713.

Burt WH (1943) Territoriality and home range concepts as applied to mammals. *Journal of Mammalogy* **24**, 346–352.

Caplan AI (1991) Mesenchymal stem cells. *Journal of Orthopaedic Research* **9**, 641–650.

Carbone C, Mace GM, Roberts SC and Macdonald DW (1999) Energetic constraints on the diet of terrestrial carnivores. *Nature* **402**, 286–288.

Carmichael LE, Krizan J, Nagy JA, Dumond M, Johnson D, Veitch A and Strobeck C (2008) Northwest passages: conservation genetics of Arctic Island wolves. *Conservation Genetics* **9**, 879–892.

Carr GM and Macdonald DW (1986) The sociality of solitary foragers: a model based on resource dispersion. *Animal Behaviour* **34**, 1540–1549.

Catling PC (1979) Seasonal variation in plasma testosterone and the testis in captive male dingoes, *Canis familiaris dingo. Australian Journal of Zoology* **27**, 939–944.

Catling PC, Burt RJ and Forrester RI (2000) Models of the distribution and abundance of ground-dwelling mammals in the eucalypt forests of north-eastern New South Wales in relation to habitat variables. *Wildlife Research* **27**, 639–654.

Catling PC, Corbett LK, Westcott M (1991) Age determination in the dingo and crossbreeds. *Wildlife Research* **18**, 75–83.

Catling PC, Reid A and Claridge AW (2005) *The Sand Plot Technique: Sampling Ground-Dwelling Vertebrates*. NSW Department of Environment and Conservation and CSIRO Sustainable Ecosystems, Canberra.

Cavilli-Sforza LL and Feldman MW (1981) *Cultural Transmission and Evolution: A Quantitative Approach*. Princeton University Press, Princeton, NJ.

Chambers GK and MacAvoy ES (2000) Microsatellites: consensus and controversy. *Comparative Biochemistry and Physiology Part B* **126**, 455–476.

Choi KC and Jeung EB (2003) The biomarker and endocrine disruptors in mammals. *Journal of Reproduction and Development* **49**, 337–345.

Ciucci P, Boitani L, Francisci F and Andreoli G (1997) Home range, activity and movements of a wolf pack in central Italy. *Journal of Zoology* **243**, 803–819.

Claridge AW and Hunt R (2008) Evaluating the role of the dingo as a trophic regulator: additional practical suggestions. *Ecological Management and Restoratrion* **9**, 116–119.

Claridge AW and Mills DJ (2007) Aerial baiting for wild dogs has no observable impact on spotted-tailed quolls (*Dasyurus maculatus*) in a rainshadow woodland. *Wildlife Research* **34**, 116–124.

Claridge AW, Murray AJ, Dawson J, Poore R, Mifsud G and Saxon MJ (2006) The propensity of spotted-tailed quolls (*Dasyurus maculatus*) to encounter and consume non-toxic meat baits in a simulated canid control program. *Wildlife Research* **33**, 85–91.

Coelho CM, Bandeira de Melo LF, Lima Sábato MA, Rizel DN and Young RJ (2007) A note on the use of GPS collars to monitor wild maned wolves *Chrysocyon brachyurus* (Illiger 1815) (Mammalia, Canidae). *Applied Animal Behaviour Science* **105**, 259–264.

Colborn T, Saal FSV and Soto AM (1993) Developmental effects of endocrine-disrupting chemicals in wildlife and humans. *Environmental Health Perspectives* **101**, 378–384.

Corbett LK (1988) Social dynamics of a captive dingo pack: population regulation by dominant female infanticide. *Ethology* **78**, 177–198.

Corbett LK (1995) Does dingo predation or buffalo competition regulate feral pig populations in the Australian wet-dry tropics? An experimental study. *Wildlife Research* **22**, 65–74.

Corbett LK (2001) *The Dingo in Australia and Asia*. J.B. Books, Marleston, SA.

Corbett LK and Newsome AE (1987) The feeding ecology of the dingo. III. Dietary relationships with widely fluctuating prey populations in arid Australia: an hypothesis of alternation of predation. *Oecologia* **74**, 215–227.

Craighead JJ, Frank C, Craighead J, Ruff RL and O'Gara BW (1973) Home range and activity patterns of nonmigratory elk of the Madison Drainage herd as determined by biotelemetry. *Wildlife Monographs* **33**, 1–50.

Creel SR and Macdonald DW (1995) Sociality, group size, and reproductive suppression among carnivores. *Advances in the Study of Behaviour* **24**, 203–257.

Crews D, Willingham E and Skipper JK (2000) Endocrine disruptors: present issues, future directions. *The Quarterly Review of Biology* **75**, 243–260.

Crooks KR and Soulé ME (1999) Mesopredator release and avifaunal extinctions in a fragmented system. *Nature* **400**, 563–566.

Curtis K (2009) 'Recent Changes in the Australian Sheep Industry (The Disappearing Flock)'. Department of Agriculture and Food, Western Australia.

D'Eon RG, Serrouya R, Smith G and Kochanny CO (2002) GPS telemetry error and bias in mountainous terrain. *Wildlife Society Bulletin* **30**, 430–439.

Daniels MJ and Corbett LK (2003) Redefining introgressed protected mammals: when is a wildcat a wild cat and a dingo a wild dog? *Wildlife Research* **30**, 213–218.

Davis E (2001) Legislative issues relating to control of dingoes and other wild dogs in New South Wales I. Approaches to future management. In: *A Symposium on the Dingo.* (Eds CR Dickman and D Lunney) Royal Zoological Society of New South Wales, Mosman.

Di Orio AP, Callas R and Schaefer RJ (2003) Performance of two GPS telemetry collars under different habitat conditions. *Wildlife Society Bulletin* **31**, 372–379.

Dickman C, Glen A and Letnic M (2009) Reintroducing the dingo: can Australia's conservation wastelands be restored? In: *Reintroduction of Top-order Predators.* (Eds MW Hayward and MJ Somers) Blackwell Publishing, Oxford.

Dickman CR and Lunney D (Eds) (2001) *A Symposium on the Dingo.* Royal Zoological Society of New South Wales, Mosman.

Edgar JP, Appleby RG and Jones DN (2007) Efficacy of an ultrasonic device as a deterrent to dingoes (*Canis lupus dingo*): a preliminary investigation. *Journal of Ethology* **25**, 209–213.

Edwards MG, de Prue ND, Shakeshaft BJ and Crealy IV (2000) An evaluation of two methods of assessing feral cat and dingo abundance in central Australia. *Wildlife Research* **27**, 143–149.

Eldridge SR, Berman DM and Walsh B (2000) Field evaluation of four 1080 baits for dingo control. *Wildlife Research* **27**, 495–500.

Eldridge SR, Shakeshaft BJ and Nano TJ (2002) 'The impact of wild dog control on cattle, native and introduced herbivores and introduced predators in central Australia'. Unpublished, Final administrative report to the Bureau of Rural Sciences.

Elledge AE, Leung LK-P, Allen LR, Firestone K and Wilton AN (2006) Assessing the taxonomic status of dingoes *Canis familiaris dingo* for conservation. *Mammal Review,* **36**, 142–156.

Evanno G, Regnaut S and Goudet J (2005) Detecting the number of clusters of individuals using the software *STRUCTURE*: a simulation study. *Molecular Ecology* **14**, 2611–2620.

Fahey J, O'Sullivan K, Crilly J and Mee JF (2002) The effect of feeding and management practices on calving rate in dairy herds *Animal Reproduction Science* **74**, 133–150.

Falush D, Stephens M and Pritchard JK (2003) Inference of population structure using multilocus genotype data: linked loci and correlated allele frequencies. *Genetics* **164**, 1567–1587.

Falush D, Stephens M and Pritchard JK (2007) Inference of population structure using multilocus genotype data: dominant markers and null alleles. *Molecular Ecology Notes* **7**, 574–578.

Faubet P, Waples RS and Gaggiotti OE (2007) Evaluating the performance of a multilocus Bayesian method for the estimation of migration rates. *Molecular Ecology* 16, 1149–1166.

Fitzgerald G and Wilkinson R (2009) 'Assessing the social impact of invasive animals in Australia'. Invasive Animals Cooperative Research Centre, Canberra.

Fleming PJS (2001) Legislative issues relating to control of dingoes and other wild dogs in New South Wales II: historical and technical justifications for current policy. In: *A Symposium on the Dingo*. (Eds CR Dickman and D Lunney) Royal Zoological Society of New South Wales, Mosman.

Fleming PJS, Allen LR, Berghout MJ, Meek PD, Pavlov PM, Stevens PL, Strong K, Thompson JA and Thomson PC (1998) The performance of wild-canid traps in Australia: efficiency, selectivity and trap-related injuries. *Wildlife Research* 25, 327–338.

Fleming PJS, Allen LR, Lapidge SJ, Robley A, Saunders GR and Thomson PC (2006) A strategic approach to mitigating the impacts of wild canids: proposed activities of the Invasive Animals Cooperative Research Centre. *Australian Journal of Experimental Agriculture* 46, 753–762.

Fleming PJS, Corbett LK, Harden R and Thomson PC (2001) *Managing the impacts of dingoes and other wild dogs*. Bureau of Rural Sciences, Canberra.

Fleming PJS, Thompson JA and Nicol HI (1996) Indices for measuring the efficacy of aerial baiting for wild dog control in north-eastern New South Wales. *Wildlife Research* 23, 665–674.

Foran B, Lenzen M and Dey C (2005) *Balancing Act: A Triple Bottom Line Analysis of the Australian Economy.* Department of the Environment and Heritage, Canberra.

Francisco LV, Langston AA, Mellersh CS, Neal CL and Ostrander EA (1996) A class of highly polymorphic tetranucleotide repeats for canine genetic mapping. *Mammalian Genome* 7, 359–362.

Frank H (1980) Evolution of canine information processing under conditions of natural and artificial selection. *Zeitschrift für Tierpsychologie* 53, 389–399.

Freeman C (2005) Is this picture worth a thousand words? An analysis of Harry Burrell's photograph of a thylacine with a chicken. *Australian Zoologist* 33, 1–16.

Friend GR (1978) A comparison of predator scat analysis with conventional techniques in a mammal survey of contrasting habitats in Gippsland, Victoria. *Australian Wildlife Research* 5, 75–83.

Gantz GF and Knowlton FF (2005) Seasonal activity areas of coyotes in the Bear River Mountains of Utah and Idaho. *Journal of Wildlife Management* 69, 1652–1659.

Geffen E, Gompper ME, Gittleman JL, Luh H-K, MacDonald DW and Wayne RK (1996) Size, life-history traits, and social organization in the Canidae: a re-evaluation. *The American Naturalist* 147, 140–160.

Gese EM, Ruff RL and Crabtree RL (1996) Social and nutritional factors influencing the dispersal of resident coyotes. *Animal Behaviour* 52, 1025–1043.

Gilbert SF (2002) The genome in its ecological context: philosophical perspectives on interspecies epigenesis. *Annals New York Academy of Sciences* 981, 202–218.

Ginsberg JR and Macdonald DW (1990) *Foxes, Wolves, Jackals, and Dogs, the IUCN/SSC Canid Specialist Group's 1990 Action Plan.* IUCN/SSC Canid Specialist Group, IUCN Publications, Gland, Switzerland.

Girman DJ, Mills MGL, Geffen E and Wayne RK (1997) A molecular genetic analysis of social structure, dispersal, and interpack relationships of the African wild dog (*Lycaon pictus*). *Behavioural Ecology and Sociobiology* **40**, 187–198.

Glen AS and Dickman CR (2005) Complex interactions among mammalian carnivores in Australia, and their implications for wildlife management *Biological Review* **80**, 1–15.

Glen AS, Dickman CR, Soulé ME and Mackey BG (2007) Evaluating the role of the dingo as a trophic regulator in Australian ecosystems. *Austral Ecology* **32**, 492–501.

Goldberg AD, Allis CD and Bernstein E (2007) Epigenetics: a landscape takes shape. *Cell,* **128**, 635–638.

Gong W, Sinden J, Braysher M and Jones R (2009) 'The economic impacts of vertebrate pests in Australia'. Invasive Animals Cooperative Research Centre, Canberra.

Goudet J (1995) FSTAT (vers. 1.2): a computer program to calculate F-statistics. *Journal of Heredity* **86**, 485–486.

Graves TA and Waller JS (2006) Understanding the causes of missed global positioning system telemetry fixes. *Journal of Wildlife Management* **70**, 844–851.

Green BF and Catling PC (1977) The biology of the dingo. In: *Australian Animals and their Environment*. (Eds H Messel and ST Butler) Shakespeare Head Press, Sydney.

Green K (2002) Selective predation on the broad-toothed rat, *Mastacomys fuscus* (Rodentia: Muridae), by the introduced red fox, *Vulpes vulpes* (Carnivora: Canidae), in the Snowy Mountains, Australia. *Austral Ecology* **27**, 353–359.

Green K and Osborne WS (1981) The diet of foxes, *Vulpes vulpes* (L.), in relation to abundance of prey above the winter snowline in New South Wales. *Australian Wildlife Research* **8**, 349–360.

Griffiths T (2001) One hundred years of environmental crisis. *Australian Rangelands Journal* **23**, 5–14.

Hall SA and Paruelo JM (2006) Environmental controls on lambing rate in Patagonia (Argentina): a regional approach. *Journal of Arid Environments* **64**, 713–735.

Harden RH (1985) The ecology of the dingo in north-eastern New South Wales I: movements and home range. *Australian Wildlife Research* **12**, 25–37.

Hatton TJ, Pierce LL and Source JW (1993) Ecohydrological changes in the Murray-Darling Basin II: development and tests of a water balance model. *Journal of Applied Ecology* **30**, 274–282.

Hemson G, Johnson P, South A, Kenward R, Ripley R and McDonald D (2005) Are kernels the mustard? Data from global positioning system (GPS) collars suggests problems for kernel home-range analyses with least-squares cross-validation. *Journal of Animal Ecology* **74**, 455–463.

Henikoff S and Matzke MA (1997) Exploring and explaining epigenetic effects. *Trends in Genetics* **13**, 293–295.

Herring SW (1993) Formation of the vertebrate face: epigenetic and functional influences. *American Zoologist* **33**, 472–483.

Hewitt L (2009) 'Major Economic Costs Associated with Wild Dogs in the Queensland Grazing Industry'. Blueprint for the Bush, AgForce, Queensland.

Heyes CM and Ray ED (2000) What is the significance of imitation in animals? *Advances in the Study of Behaviour* **29**, 215–245.

Holmes NG, Dickens HF, Parker HL, Binns MM, Mellersh CS and Sampson J (1995) Eighteen canine microsatellites. *Animal Genetics* **26**, 132–133.

Holmes NG, Mellersh CS, Humphreys SJ, Binns MM, Holliman J, Curtis R and Sampson J (1993) Isolation and characterization of microsatellites from the canine genome. *Animal Genetics* **24**, 289–292.

Hulst F (2008) Dingoes. In: *Medicine of Australian Mammals*. (Eds L Vogelnest and R Woods) CSIRO Publishing, Collingwood.

Hytten KF (2009) Dingo dualisms: exploring the ambiguous identity of Australian dingoes. *Australian Zoologist* **35**, 18–27.

Johnson CN and VanDerWal J (2009) Evidence that dingoes limit abundance of a mesopredator in eastern Australian forests. *Journal of Applied Ecology* **46**, 641–646.

Johnson CN and Wroe S (2003) Causes of extinction of vertebrates during the Holocene of mainland Australia: arrival of the dingo, or human impact? *The Holocene* **13**, 941–948.

Johnson CN, Isaac JL and Fisher DO (2006) Rarity of a top predator triggers continent-wide collapse of mammal prey: dingoes and marsupials in Australia. *Proceedings of the Royal Society B: Biological Sciences* **274**, 341–346.

Jones E (1990) Physical characteristics and taxonomic status of wild canids, *Canis familiaris*, from the eastern highlands of Victoria. *Australian Wildlife Research* **17**, 69–81.

Jones E (2009) Hybridisation between the dingo, *Canis lupus dingo*, and the domestic dog, *Canis lupus familiaris*, in Victoria: a critical review. *Australian Mammalogy* **31**, 1–7.

Jones E and Stevens PL (1988) Reproduction in wild canids, *Canis familiaris*, in the eastern highlands of Victoria. *Australian Wildlife Research* **15**, 385–394.

Kelce WR, Gray LE and Wilson EM (1998) Antiandrogens as environmental endocrine disruptors. *Reproduction, Fertility and Development* **10**, 105–111.

Kenward RE (2001) *A Manual for Wildlife Radio Tagging*. Academic Press, London.

Kiernan K, Jones R and Ranson D (1983) New evidence from Fraser Cave for glacial age man in south-west Tasmania. *Nature* **301**, 28–32.

Kindall JL and Van Manen FT (2007) Identifying habitat linkages for American black bears in North Carolina, USA. *Journal of Wildlife Management* **71**, 487–495.

Kleiman D (1968) Reproduction in the Canidae. *International Zoo Yearbook* **8**, 3–8.

Kohen J (1993) *The Darug and their neighbours: the traditional Aboriginal owners of the Sydney region*. Darug Link in association with Blacktown and District Historical Society, Blacktown, NSW.

Kohen JL (1995) *Aboriginal Environmental Impacts*. University of New South Wales Press, Sydney.

Körtner G and Watson P (2005) The immediate impact of 1080 aerial baiting to control wild dogs on a spotted-tailed quoll population. *Wildlife Research* **32**, 673–680.

Koskinen MT and Bredbacka P (2000) Assessment of the population structure of five Finnish dog breeds with microsatellites. *Animal Genetics* **31**, 310–317.

Laundré JW and Keller BL (1981) Home-range use by coyotes in Idaho. *Animal Behaviour* **29**, 449–461.

Lehman N, Clarkson P, Mech LD, Meier TJ and Wayne RK (1992) A study of the genetic relationships within and among wolf packs using DNA fingerprinting and mitochondrial DNA. *Behavioural Ecology and Sociobiology* **30**, 83–94.

Leonard JA, Wayne RK, Wheeler J, Valadez R, Guillén S and Vilà C (2002) Ancient DNA evidence for old world origin of new world dogs. *Science* **298**, 1613–1616.

Lindström E (1986) Territory inheritance and the evolution of group-living in carnivores. *Animal Behaviour* **34**, 1825–1835.

Lunney D (2001) Causes of the extinction of native mammals of the western division of New South Wales: An ecological interpretation of the nineteenth century historical record. *Australian Rangelands Journal* **23**, 44–70.

Lunney D, Triggs B, Eby P and Ashby E (1990) Analysis of scats of dogs *Canis familiaris* and foxes *Vulpes vulpes* (Canidae: Carnivora) in coastal forests near Bega, New South Wales. *Australian Wildlife Research* **17**, 61–68.

Lyon MF (1993) Epigenetic inheritance in mammals. *Trends in Genetics* **9**, 123–128.

Macdonald DW (1983) The ecology of carnivore social behaviour. *Nature* **301**, 379–384.

Macdonald DW and Sillero-Zubiri C (Eds) *Biology and Conservation of Wild Canids*. Oxford University Press, Oxford.

Macdonald DW, Creel S and Mills MGL (2004) Society. In: *Biology and Conservation of Wild Canids*. (Eds DW Macdonald and C Sillero-Zubiri) Oxford University Press, Oxford.

Macdonald DW and Sillero-Zubiri C (2004a) Conservation. In: *Biology and Conservation of Wild Canids*. (Eds DW Macdonald and C Sillero-Zubiri) Oxford University Press, Oxford.

Macdonald DW and Sillero-Zubiri C (2004b) *Dramatis personae*. In: *Biology and Conservation of Wild Canid*. (Eds DW Macdonald and C Sillero-Zubiri) Oxford University Press, Oxford.

Macintosh NWG (1976) The origin of the dingo: an enigma. In: *The Wild Canids: Their Systematics, Behavioural Ecology and Evolution*. (Ed. MW Fox) Dogwise Publishing, Washington.

Marcus J (1989) Prisoner of discourse: the dingo, the dog and the baby. *Anthropology Today* **5**, 15–19.

Marsack P and Campbell G (1990) Feeding behaviour and diet of dingoes in the Nullarbor Region, Western Australia. *Australian Wildlife Research* **17**, 349–357.

May SA and Norton TW (1996) Influence of fragmentation and disturbance on the potential impact of feral predators on native fauna in Australian forest ecosystems. *Wildlife Research* **23**, 387–400.

Mech LD (1970) *The Wolf: The Ecology and Behaviour of an Endangered Species*. Natural History Press, New York.

Mech LD and Barber SM (2002) 'A critique of wildlife radio-tracking and its use in national parks'. Northern Prairie Wildlife Research Centre and University of Minnesota, Jamestown.

Meek PD (1999) The movement, roaming behaviour and home range of free-roaming domestic dogs, *Canis lupus familiaris*, in coastal New South Wales. *Wildlife Research* **26**, 847–856.

Meek PD and Triggs B (1998) The food of foxes, dogs and cats on two peninsulas in Jervis Bay, New South Wales. *Proceedings of the Linnean Society NSW* **120**, 117–127.

Meek PD, Jenkins DJ, Morris B, Ardler AJ and Hawksby RJ (1995) Use of two human leg-hold traps for catching pest species. *Wildlife Research* **22**, 733–739.

Mellersh C, Holmes N, Binn M and Sampson J (1994) Dinucleotide repeat polymorphisms at four canine loci (LEI 003, LEI 007, LEI 008 and LEI 015). *Animal Genetics* **25**, 125.

Mellersh CS, Langston AA, Acland GM, Fleming MA, Ray K, Wiegand NA, Francisco LV, Gibbs M, Aguirre GD and Ostrander EA (1997) A linkage map of the canine genome. *Genomics* **46**, 326–336.

Merrill SB and Mech LD (2003) The usefulness of GPS telemetry to study wolf circadian and social activity. *Wildlife Society Bulletin* **31**, 947–960.

Messier F and Barrette C (1982) The social system of the coyote (*Canis latrans*) in a forested habitat. *Canadian Journal of Zoology* **60**, 1743–1753.

Mitchell BD and Banks PB (2005) Do wild dogs exclude foxes? Evidence for competition from dietary and spatial overlaps. *Austral Ecology* **30**, 581–591.

Moehlman PD (1986) Ecology of cooperation in canids. In: *Ecological Aspects of Social Evolution: Birds and Mammals.* (Eds DI Rubinstein and RW Wrangham) Princeton University Press, Princeton, NJ.

Moehlman PD (1989) Intraspecific variation in canid social systems. In: *Carnivore Behaviour, Ecology and Evolution.* (Ed. JL Gittleman) Cornell University Press, Ithaca, NY.

Moehlman PD and Hofer H (1997) Cooperative breeding, reproductive suppression and body mass in canids. In: *Cooperative Breeding in Mammals.* (Eds NG Solomon and JA French) Cambridge University Press, Cambridge, UK.

Monk M (1990) Variation in epigenetic inheritance. *Trends in Genetics* **6**, 110–114.

Montagu MFA (1942) On the origin of the domestication of the dog. *Science* **96**, 111–112.

Morey DF (1992) Size, shape and development in the evolution of the domestic dog. *Journal of Archaelogical Science* **19**, 181–204.

Nakada M and Lambeck K (1989) Late Pleistocene and Holocene sea-level change in the Australian region and mantle rheology. *Geophysical Journal* **96**, 497–517.

Nesbitt B, Wilton AN and Jenkins DJ (2000) Dingoes of the New England and Guy Fawkes national parks of north east New South Wales – are there any left out there? In: *NSW Pest Animal Control Conference – Practical Solutions to Pest Management Problems.* (Ed. S Balough) NSW Agriculture, Orange, NSW.

Nesbitt WH (1975) Ecology of a feral dog pack on a wildlife refuge. In: *The Wild Canids*. (Ed. MW Fox) Van Nostrand Reinhold, New York.

Newsome AE (2001) The biology and ecology of the dingo. In: *A Symposium on the Dingo*. (Eds CR Dickman and D Lunney) The Royal Zoological Society of New South Wales, Mosman.

Newsome AE and Coman BJ (1989) Canidae. In: *Fauna of Australia. Mammalia*. (Eds DW Dalton and BJ Richardson) Australian Government Publishing Service, Canberra.

Newsome AE and Corbett LK (1982) The identity of the dingo II.* Hybridization with domestic dogs in captivity and in the wild. *Australian Journal of Zoology* **30**, 365–374.

Newsome AE and Corbett LK (1985) The identity of the dingo III.* The incidence of dingoes, dogs and hybrids and their coat colours in remote and settled regions of Australia. *Australian Journal of Zoology* **33**, 363–375.

Newsome AE, Catling PC and Corbett LK (1983a) The feeding ecology of the dingo II. Dietary and numerical relationships with fluctuating prey populations in south-eastern Australia. *Australian Journal of Ecology* **8**, 345–366.

Newsome AE, Corbett LK and Carpenter SM (1980) The identity of the dingo I. Morphological discriminants of dingo and dog skulls. *Australian Journal of Zoology* **28**, 615–625.

Newsome AE, Corbett LK, Catling PC and Burt RJ (1983b) The feeding ecology of the dingo. I. Stomach contents from trapping in south-eastern Australia, and non-target wildlife also caught in dingo traps. *Australian Wildlife Research* **10**, 477–486.

Oakman B (2001) The problem with keeping dingoes as pets and conservation. In: *A Symposium on the Dingo*. (Eds CR Dickman and D Lunney) The Royal Zoological Society of New South Wales, Mosman.

Ostrander EA, Sprague Jr, GF and Rine J (1993b) Identification and characterization of dinucleotide repeat $(CA)_n$ markers for genetic mapping in dog. *Genomics* **16**, 207–213.

Paltridge R (2002) The diet of cats, foxes and dingoes in relation to prey availability in the Tanami Desert, Northern Territory. *Wildlife Research* **29**, 389–403.

Paplinska JZ, Moyle RLC, Temple-Smith PDM and Renfree MB (2006) Reproduction in female swamp wallabies, *Wallabia bicolor*. *Reproduction, Fertility and Development* **18**, 735–743.

Parker HG, Kim LV, Sutter NB, Carlson S, Lorentzen TD, Malek TB, Johnson GS, DeFrance HB, Ostrander EA and Kruglyak L (2004) Genetic structure of the purebred domestic dog. *Science* **304**, 1160–1163.

Parker M (2007) The cunning dingo. *Society and Animals* **15**, 69–78.

Parker MA (2006) Bringing the dingo home: discursive representations of the dingo by Aboriginal, colonial and contemporary Australians. PhD Thesis, University of Tasmania.

Pierce LL, Walker J, Dowling TI, McVicar TR, Hatton TJ, Running SW and Coughlan JC (1993) Ecohydrological changes in the Murray-Darling Basin. III. A simulation of regional hydrological changes. *Journal of Applied Ecology* **30**, 283–294.

Pink B (2008) *Water and the Murray-Darling Basin – A Statistical Profile Australia 2000–01 to 2005–06*. Australian Bureau of Statistics, Canberra.

Pongrácz P, Miklósi Á, Kubinyi E, Topáli J and Csányi V (2003) Interaction between individual experience and social learning in dogs. *Animal Behaviour* **65**, 595–603.

Pritchard JK, Stephens M and Donnelly P (2000a) Inference of population structure using multilocus genotype data. *Genetics* **155**, 945–959.

Pritchard JK, Stephens M, Rosenberg NA and Donnelly P (2000b) Association mapping in structured populations. *American Journal of Human Genetics* **67**, 170–181.

Purcell BV (2010) Order in the pack: ecology of *Canis lupus dingo* in the Southern Greater Blue Mountains World Heritage Area. PhD Thesis, University of Western Sydney, Richmond.

Purcell BV, Mulley R, Close R and Fleming P (2006) Use of GPS collars for tracking wild dogs. *Queensland Pest Animal Symposium Proceedings*, Toowoomba.

Quinn M (2001) Rights to the rangelands: European contests of possession in the early 20th century. *Australian Rangelands Journal* **23**, 5–14.

Rakyan V and Whitelaw E (2003) Transgenerational epigenetic inheritance. *Current Biology* **13**(1), R6.

Rakyan VK and Beck S (2006) Epigenetic variation and inheritance in mammals. *Current Opinion in Genetics and Development* **16**, 573–577.

Ridley J, Yu DW and Sutherland WJ (2004) Why long-lived species are more likely to be social: the role of local dominance. *Behavioural Ecology* **16**, 358–363.

Ripple WJ and Beschta RL (2003) Wolf reintroduction, predation risk, and cottonwood recovery in Yellowstone National Park. *Forest Ecology and Management* **184**, 299–313.

Ritchie EG and Johnson CN (2009) Predator interactions, mesopredator release and biodiversity conservation. *Ecology Letters* **12**, 1–18.

Robertshaw JD and Harden RH (1985) The ecology of the dingo in north-eastern New South Wales II. Diet. *Australian Wildlife Research* **12**, 39–50.

Robertshaw JD and Harden RH (1986) The ecology of the dingo in north-eastern New South Wales IV. Prey selection by dingoes, and its effect on the major prey species, the swamp wallaby, *Wallabia bicolor* (Desmarest). *Australian Wildlife Research* **13**, 141–163.

Roemer I, Reik W, Dean W and Klose J (1997) Epigenetic inheritance in the mouse. *Current Biology* **7**, 277–280.

Ruvinsky A (2001) Developmental Genetics. In: *The Genetics of the Dog*. (Eds A Ruvinsky and J Sampson) CABI Publishing, Wallingford.

Samuel MD and Green RE (1988) A revised test procedure for identifying core areas within the home range. *Journal of Animal Ecology* **57**, 1067–1068.

Savolainen P, Leitner T, Wilton AN, Matisoo-Smith E and Lundeberg J (2004) A detailed picture of the origin of the Australian dingo, obtained from the study of mitochondrial DNA. *Proceedings of the National Academy of Sciences of the United States of America* **101**, 12387–12390.

Savolainen P, Zhang Y, Luo J, Lundeberg J and Leitner T (2002) Genetic evidence for an east Asian origin of domestic dogs. *Science* **298**, 1610–1613.

Scott JP (1968) Evolution and domestication of the dog. *Evolutionary Biology* **2**, 243–275.

Shepherd NC (1981) Predation of red kangaroos, *Macropus rufus*, by the dingo, *Canis familiaris dingo* (Blumenbach), in north-western New South Wales. *Australian Wildlife Research* **8**, 255–262.

Sillero-Zubiri C and Macdonald DW (1998) Scent-marking and territorial behaviour of Ethiopian wolves *Canis simensis*. *Journal of Zoology* **245**, 351–361.

Sillero-Zubiri C, Hoffman M and Macdonald DW (2004) *Canids: Foxes, Wolves, Jackals and Dogs, Status Survey and Conservation Action Plan*. IUCN/SSC Canid Specialist Group. Gland, Switzerland and Cambridge, UK.

Sillero-Zubiri C, Reynolds J and Novaro AJ (2004) Management. In: *Biology and Conservation of Wild Canids*. (Eds DW Macdonald and C Sillero-Zubiri) Oxford University Press, Oxford.

Slabbert JM and Rasa OAE (1997) Observational learning an acquired maternal behaviour pattern by working dog pups: an alternative training method. *Applied Animal Behaviour Science* **53**, 309–316.

Sponenberg DP and Rothschild MF (2001) Genetics of coat colour and hair texture. In: *The Genetics of the Dog*. (Eds A Ruvinsky and J Sampson) CABI Publishing, Wallingford.

Steinfield H, Gerber P, Wassenaar T, Catel V, Rosales M and de Haan C (2006) *Livestock's long shadow: environmental issues and options*. Food and Agriculture Organization of the United Nations, Rome.

Thompson JA, Fleming PJS and Heap EW (1990) The accuracy of aerial baiting for wild dog control in New South Wales. *Australian Wildlife Research* **17**, 209–217.

Thompson PG (1997) The public health impact of dog attacks in a major Australian city. *The Medical Journal of Australia* **167**, 129–132.

Thomson PC (1986) The effectiveness of aerial baiting for the control of dingoes in north-western Australia. *Australian Wildlife Research* **13**, 165–176.

Thomson PC (1992a) The behavioural ecology of dingoes in north-western Australia. I. The Fortescue River study area and details of captured dingoes. *Wildlife Research* **19**, 509–518.

Thomson PC (1992b) The behavioural ecology of dingoes in north-western Australia. II. Activity patterns, breeding season and pup rearing. *Wildlife Research* **19**, 519–530.

Thomson PC (1992c) The behavioural ecology of dingoes in north-western Australia. III. Hunting and feeding behaviour, and diet. *Wildlife Research* **19**, 531–541.

Thomson PC (1992d) The behavioural ecology of dingoes in north-western Australia. IV. Social and spatial organisation, and movements. *Wildlife Research* **19**, 543–563.

Thomson PC and Rose K (1992) Age determination of dingoes from characteristics of canine teeth. *Wildlife Research* **19**, 597–599.

Thomson PC, Rose K and Kok NE (1992a) The behavioural ecology of dingoes in north-western Australia. V. Population dynamics and variation in the social system. *Wildlife Research* **19**, 565–584.

Thomson PC, Rose K and Kok NE (1992b) The behavioural ecology of dingoes in north-western Australia. VI. Temporary extraterritorial movements and dispersal. *Wildlife Research* **19**, 585–595.

Triggs B (1996) *Tracks, Scats and Other Traces: A Field Guide to Australian Mammals.* Oxford University Press, Melbourne.

Trut LN (2001) Experimental studies of early canid domestication. In: *The Genetics of the Dog.* (Eds A Ruvinsky and J Sampson) CABI Publishing, Wallingford.

Trut LN, Plyusnina IZ and Oskina IN (2004) An experiment on fox domestication and debatable issues of evolution of the dog. *Russian Journal of Genetics* **40**, 644–655.

Twigg LE, Eldridge SR, Edwards GP, Shakeshaft BJ, dePrue ND and Adams N (2000) The longevity and efficacy of 1080 meat baits used for dingo control in central Australia. *Wildlife Research* **27**, 473–481.

Van Ballenberghe V (1983) Two litters raised in one year by a wolf pack. *Journal of Mammalogy* **64**, 171–173.

Van Speybroeck L, De Waele D and Van De Vijver G (2002) Theories in early embryology: close connections between epigenesis, preformationism, and self-organization. *Annals of the New York Academy of Sciences* **981**, 7–49.

Van Valkenburgh B (2007) Déjà vu: the evolution of feeding morphologies in the Carnivora. *Integrative and Comparative Biology* **47**, 147–163.

Vilà C, Savolainen P, Maldonado JE, Amorim IR, Rice JE, Honeycutt RL, Crandall KA, Lundeberg J and Wayne RK (1997) Multiple and ancient origins of the domestic dog. *Science* **276**, 1687–1689.

vonHoldt BM, Pollinger JP, Lohmueller KE, Han E, Parker HG, Quignon P, Degenhardt JD, Boyko AR, Earl DA, Auton A, Reynolds A, Bryc K, Brisbin A, Knowles JC, Mosher DS, Spady TC, Elkahloun A, Geffen E, Pilot M, Jedrzejewski W, Greco C, Randi E, Bannasch D, Wilton A, Shearman J, Musiani M, Cargill M, Jones PG, Qian Z, Huang W, Ding Z-L, Zhang Y, Bustamante CD, Ostrander EA, Novembre J and Wayne RK (2010) Genome-wide SNP and haplotype analyses reveal a rich history underlying dog domestication. *Nature* **464**, 898–902.

Walker J, Bullen F and Williams BG (1993) Ecohydrological changes in the Murray-Darling Basin. I. The number of trees cleared over two centuries. *Journal of Applied Ecology* **30**, 265–273.

Wallach AD and O'Neill AJ (2008) 'Persistence of endangered species: is the dingo the key?' Final report to Department of Environment and Heritage, Adelaide.

Wallach AD, Murray B and O'Neill AJ (2008) Can threatened species survive where the top predator is absent? *Biological Conservation* **142**, 43–52.

Wallach AD, Ritchie EG, Read J and O'Neill AJ (2009) More than mere numbers: the impact of lethal control on the social stability of a top order predator. *PLoS ONE* **4**, e6861.

Wang X, Tedford RH, Van Valkenburgh B and Wayne RK (2004) Ancestry. In: *Biology and Conservation of Wild Canids.* (Eds DW Macdonald and C Sillero-Zubiri) Oxford University Press, Oxford.

Wayne RK (1986) Cranial morphology of domestic and wild canids: the influence of development on morphological change. *Evolution* **40**, 243–261.

Wayne RK (2001) Consequences of domestication: morphological diversity of the dog. In: *The Genetics of the Dog*. (Eds A Ruvinsky and J Sampson) CABI Publishing, Wallingford.

Wayne RK and Ostrander EA (1999) Origin, genetic diversity, and genome structure of the domestic dog. *BioEssays* **21**, 247–257.

Wayne RK and Ostrander EA (2007) Lessons learned from the dog genome. *Trends in Genetics* **23**, 557–567.

Wayne RK and Vilà C (2001) Phylogeny and origin of the domestic dog. In: *The Genetics of the Dog*. (Eds A Ruvinsky and J Sampson) CABI Publishing, Wallingford.

Wayne RK, Geffen E, Girmin DJ, Koepfli KP, Lau LM and Marshall C (1997) Molecular systematics of the Canidae. *Systematic Biology* **4**, 622–653.

Wayne RK, Geffen E and Vilà C (2004) Population genetics. In: *Biology and Conservation of Wild Canids*. (Eds DW Macdonald and C Sillero-Zubiri) Oxford University Press, Oxford.

White PJ and Garrott RA (1995) Northern Yellowstone elk after wolf restoration. *Wildlife Society Bulletin* **33**, 942–955.

Whitehouse SJO (1977) The diet of the dingo in Western Australia. *Australian Wildlife Research* **4**, 145–150.

Wilcox C and Curtis K (2009) 'Situation, outlook and opportunities for the supply and demand of apparel wool'. Poimena Analysis, Cooperative Research Centre for Sheep Industry Innovation, Armidale.

Wilton AN (2001) DNA methods of assessing dingo purity. In: *A Symposium on the Dingo*. (Eds CR Dickman and D Lunney) Royal Zoological Society of New South Wales, Mosman.

Wilton AN, Steward DJ and Zafiris K (1999) Microsatellite variation in the Australian dingo. *Journal of Heredity* **90**, 108–111.

Woodall PF, Pavlov P and Twyford KL (1996) Dingoes in Queensland, Australia: skull dimensions and the identity of wild canids. *Wildlife Research* **23**, 581–587.

Wroe S, Clausen P, McHenry C, Moreno K and Cunningham E (2007) Computer simulation of feeding behaviour in the thylacine and dingo as a novel test for convergence and niche overlap. *Proceedings of the Royal Society B: Biological Sciences* **274**, 2819–2828.

Yates D, Hayes G, Heffernan M and Beynon R (2003) Incidence of cryptorchidism in dogs and cats. *The Veterinary Record* **152**, 502–504.

Zacharewski T (1998) Identification and assessment of endocrine disruptors: limitations of *in vivo* and *in vitro* assays. *Environmental Health Perspectives* **106**, 577–582.

Zrzavý J and Ricánková V (2004) Phylogeny of the recent Canidae (Mammalia, Carnivora): relative reliability and utility of morphological and molecular datasets. *Zoologica Scripta* **33**, 311–333.

INDEX

www.ingramcontent.com/pod-product-compliance
Lightning Source LLC
Chambersburg PA
CBHW041130280526
45792CB00013B/2367